age of the COSMOS

by

HAROLD S. SLUSHER, M.S., D.Sc.

Assistant Professor of Physics
University of Texas at El Paso
and
Research Associate,
Institute for Creation Research
San Diego, California

ICR Technical Monograph No. 9

Institute for Creation Research
San Diego, CA 92116

AGE OF THE COSMOS

Copyright © 1980

Institute for Creation Research
2716 Madison Avenue
San Diego, CA 92116

Library of Congress Catalog Card
Number 79-92722
ISBN 0-932766-03-X

Cataloging in Publication Data
Slusher, Harold Schultz, 1934-
  Age of the Cosmos.
   1. Cosmogony   2. Cosmology   I. Title
       523.12   79-92722

ISBN 0-932766-03-X
Printed in United States of America

THE AUTHOR

HAROLD S. SLUSHER, M.S., D.Sc.

Professor Harold S. Slusher earned his B.A. from the University of Tennessee (mathematics and physics) and his M.S. from the University of Oklahoma (physics and astronomy). In 1975 he was awarded an honorary D.Sc. degree by Indiana Christian University in recognition of his work on the time scale for the cosmos. For many years he served as Director of the University of Texas at El Paso Kidd Memorial Seismic Observatory. He is Assistant Professor on the faculty of the Department of Physics at the University of Texas at El Paso and also serves as Research Associate in Geoscience and Astronomy with the Institute for Creation Research and is Adjunct Chairman of the Department of Physical Sciences at Christian Heritage College.

Professor Slusher's fields of interest and research include solar system astronomy, terrestrial heat flow, and cosmogony. In addition to this monograph *Age of the Cosmos,* he has also written other Institute for Creation Research monographs: *A Critique of Radiometric Dating, The Origin of the Universe: An Examination of the Big Bang and Steady-State Cosmogonies, The Age of the Solar System: A Study of the Poynting-Robertson Effect and Extinction of Interplanetary Dust,* and *The Age of the Earth: A Study of the Cooling of the Earth Under the Influence of Radioactive Heat Sources.*

TABLE OF CONTENTS

Chapter I
REMARKS ABOUT
THE AGE OF
THE COSMOS

An investigation of the age of the universe and the solar system is essential in any study of the physical history of the cosmos. The evolutionist needs vast spans of time in his history of the cosmos. His hypothesis regarding the origin and development of the cosmos says that almost imperceptible changes have occurred very gradually over vast time intervals by chance processes to produce the universe we see before us. This concept, obviously, must feed on time. The importance of the study of the chronometry of physical events beginning with the origin of the universe and coming down to the present time in arriving at the truth about the past cannot be overemphasized.

Time can be sensed only by events that occur within its span. But, in the study of the cosmos, we never observe past events themselves, only the effects of these events on the region of space through which the earth travels. The astronomer tries to infer from a study of the celestial bodies, the dynamical interactions of these bodies, and the various physical processes related to these objects, the age of the universe. In all studies of the past, but particularly in chronometry, it is tremendously important to keep actual observations separate from speculative inferences. It is very easy for the two to become interlaced in the investigation of such a complex problem as chronometry, but they may be very different. Facts are true, but inferences are derived and may be either true or false or only partially true.

In much of the work done by those studying the age of things it is not always easy to separate fact from fiction because these investigators have "assumed the things they are trying to prove" and, consequently, fail many times to make clear where fact ends and myth begins. Unfortunately, it seems that many scientists have taken up the practice of trying to force science to express their philosophies as to what the universe was like in the past.

The Second Law of Thermodynamics says that all natural processes are deteriorative or degenerative. Natural processes are changing the universe in a way similar to the unwinding of a clock spring that loses organization by the ticking of the clock. An external agent must be present and active to *make* the "clock" and to *wind* it up at the beginning. It is not possible to arrive at a unique description of the past by working backwards in a situation where disorder is continually taking place. The scientific method is not so readily usable when working into the past where there are no direct observations.

The age of the cosmos has had nearly as many values as the number of people who have studied the matter. Over the years the evolutionist has pushed the origin of the universe further and further back into a remote dim past. He has used certain techniques to get these alleged ages. In recent years creationist/catastrophists who have studied these techniques have found serious flaws in them. Further, new approaches have been introduced having as little dependence on assumptions and guesswork as possible, resulting in strong evidence for a recent creation of the universe.

It is the purpose of this monograph to look at the physical indicators leading to upper limits on the age of things. The evolutionist takes the age of the universe as close to twenty billion years and the age of the solar system and the earth as around five billion years. Evidence is advanced here that these claims of huge ages are vastly in error and a young age for the cosmos is strongly suggested.

Chapter II
THE CLUSTERS
OF GALAXIES

A. Galaxy Clusters and Their Morphology

Galaxies and clusters of galaxies are objects of much concerted study today. Galaxies in clusters are bound together by gravitational forces and, thus, provide a sort of laboratory for observations of gravitational interactions of incredible amounts of matter. Galaxies never seem to occur singly. They are only found in pairs or larger aggregates, including some with hundreds and even thousands of galaxies.

G. O. Abell divides clusters of galaxies into two general types, regular and irregular. Regular clusters are almost spherically symmetric and contain possibly many thousands of members. The irregular clusters show little symmetry and they can have large or small numbers of members in them. The galaxies in a given cluster are not distributed at random, but they have a strong tendency to occur together in subgroups within the cluster. There is evidence that the clusters themselves are not uniformly distributed, but that they tend to lump together into clusters of clusters of galaxies. The earth's galaxy, the Milky Way, is a member of a galaxy cluster called the Local Group. Our cluster, along with the Virgo Cluster and other similar groupings nearby, comprise what is called the Local Super Cluster. This cluster of clusters contains 100 galaxy clusters.

Sometimes members of galaxy pairs or of clusters are joined by bridges of luminous matter. In a few cases the speed of the galaxies along the radial direction (the component of the velocity in the direction toward or away from the observer) alone are of the order of many thousand kilometers per second. It is not likely that these galaxies are gravitationally bound. In general, the mass of the galaxies that are members of a physically well isolated group or cluster seems to be considerably smaller than the mass that would be required to bind these galaxies gravitationally. Astronomers generally hold that the members of galaxy pairs or of clusters originated simultaneously. They would, therefore, seem to have originated quite recently since they would have reached their present separations quite rapidly if coming out of roughly the same region of space.

A galaxy is a collection of some hundred billion stars held together presumably by gravity. The Local Group consists of about twenty-four or so galaxies in all. It contains a typical distribution of types of galaxies. It contains three spiral galaxies, the Milky Way, Andromeda, and M33. There are four irregular galaxies including the Large and Small Magellanic Clouds. The rest of the galaxies are ellipticals, including four regular ellipticals: Two are the companions to the Andromeda Galaxy. The others are dwarf ellipticals. The Local Group is very small compared to most of the hundreds of clusters thus far observed and catalogued. An average cluster of galaxies has one or two hundred members, while the largest contains several thousand galaxies.

The nearest cluster outside of the Local Group is thought to be about 500 billion kilometers away in straight-line distance. The most distant known clusters are believed to lie around 200 times farther away, at the very edge of the observable universe. These distances are obtained by indirect methods based on assumptions difficult to prove. This should always be kept in mind when distances of astronomical objects are given. Two of the richest clusters, one in the direction of the constellation Virgo and one in Coma Berenices, are at relatively small distances and have been studied carefully.

B. Methods for Obtaining the Mass of a Cluster

It would be instructive to consider some of the methods for obtaining the masses of galaxies and, consequently, the masses of the clusters of the galaxies. There are three basic methods for estimating the masses of galaxies. We will consider only one of these methods in some detail and take a quick glance at two of the remaining methods.

1. The mass of a galaxy can be found if the gravitational attraction of the galaxy on any object can be measured. The most accurate mass determinations are for those galaxies for which a rotation curve can be measured. This is a plot of the rotation velocity

of stars in a galaxy (usually assumed to be circular orbits) as a function of the distance from the center of the galaxy. The rotation curve for these stars depends on the distribution of matter in the galaxy.

The rotation curve of a galaxy is obtained by measuring the radial velocity of many different parts of the galaxy. The measured velocity is the average over all emitting regions of the galaxy in the line of sight, and only spirals are thin enough for this to refer very effectively to one localized region. Irregular galaxies often do not show strong evidence of rotation, so most good rotation curves are for spirals.

In order to obtain the rotation velocity, the radial velocity must be corrected for the velocity of the galaxy as a whole, and the result must then be adjusted for the appropriate projection factor in the radial direction. The latter is obtained from the observed shape of the spiral on the assumption that it would be reasonably circular if seen from straight above.

There are problems in analyzing the rotation curves to obtain masses. The distance of the galaxy must be known before the correct linear scale can be found. There are cases in which large deviations from circular motion probably occur. Even if the orbits are circular, the rotation velocity of a point is sensitive to the amount of mass within a volume of space around that point, and there may be a considerable amount of mass outside the last measured point which goes essentially undetected. In this case the mass found from the rotation curve would only be a lower limit.

Suppose that the matter in a galaxy is distributed in a spherically symmetric fashion. This is an approximation to a spiral galaxy, but it can be treated easily and it shows the main points of the method. Let $M(r)$ be the mass within the sphere of radius r and $\bar{\rho}(r)$ be the average density of the matter in this sphere. Then if r_0 is the radius of the galaxy,

$$M(r) = 4/3\, \pi\, r^3\, \bar{\rho}(r) \quad \text{FOR} \quad r \leqq r_0$$

$$M(r) = 4/3\, \pi\, r_0{}^3\, \bar{\rho}(r_0) = M \quad \text{FOR} \quad r = r_0$$

where M is the total mass of the galaxy. If a star of mass m is at a distance r from the center of the galaxy, the gravitational force on it due to the entire galaxy is $F = G\, \dfrac{m\, M(r)}{r^2}$ Using the previous expressions,

$$F = 4/3\, \pi\, Gm\, \bar{\rho}(r)\, r \quad \text{FOR} \quad r \leqq r_0$$

$$F = G\, \frac{mM}{r^2} \quad \text{FOR} \quad r = r_0.$$

The force on a mass increases with distance from the center until the edge of the galaxy is reached, and then it falls off as r^{-2}. $\bar{\rho}(r)$ is not the density at r, but it is the average density of all points within radius r.

If it is assumed that all orbits are circular, then the centrifugal force on mass m is mV^2/r. Equating this to the above expressions for force

$$V^2 = \tfrac{4}{3}\pi G \bar{\rho}(r)r^2 = \frac{GM(r)}{r} \quad \text{FOR } r \leqq r_o$$

$$V^2 = \frac{GM}{r} \quad \text{FOR } r = r_o .$$

The value for M may be then found from the corresponding value of r and V.

2. A method based on the virial theorem is also used to obtain cluster masses. The virial theorem is a statistical description of the overall dynamic behavior of a large assembly of bodies and states that over a very large period of time the average kinetic energy is one-half the average potential energy of the assembly. By observing the spread in radial Doppler velocities among different galaxies in the cluster, the mean kinetic energy per unit mass may be calculated. Then by use of the virial theorem the mean potential energy per unit mass may be calculated. If a typical cluster diameter is known from the cluster's distance and from the angle it subtends in the sky, a rough estimate of the total cluster mass may be made. In other words, if the virial theorem is satisfied for a cluster, then the observed radial velocities and positions of the members can be used to find the cluster mass and the average mass per visible galaxy.

These two methods are not in very good agreement with each other concerning the general run of galaxy masses. With few exceptions, the second method disagrees with the first by very large amounts. If the virial theorem is assumed to be valid for a given cluster, or even if the cluster is only assumed to be stable (a less restrictive assumption), then the cluster mass is usually found to be so great that the average mass per galaxy is an order of magnitude or more greater than that found for similar types of galaxies by all the other methods. This result is found for all types of clusters, regular and irregular, large and small. Even the Local Group has this characteristic.

There are possible explanations for the above discrepancy, but none remotely satisfactory. It is possible that some clusters may contain enough of the giant elliptical galaxies, systems far larger and more massive than the ordinary elliptical or spirals nearby, to account for the extra mass; but this cannot be the answer in most cases. Perhaps, the virial theorem may not be assumed since a huge amount of time must elapse before the theorem could be applicable and it assumes a stable system which does not appear to be the case at all. The most popular attempt at explaining the mass deficiency is that clusters contain large amounts of unseen material in the form of intergalactic material and dwarf galaxies too faint to be seen from very far away. This would make the clusters more massive without increasing the masses of the galaxies actually observed. As remarked later there is no evidence for this, and the unseen material would have to be vastly more abundant than the material that is visible.

The obvious explanation is that the clusters should be breaking up and are not stable and, hence, the virial theorem does not apply. The time needed for complete breakup or mechanical dissolution is defined as roughly the size of the cluster divided by typical velocity of a member.

3. From the recent studies of a number of galaxies, it is found that the total mass of a galaxy is very closely related to its brightness. Thus, it is possible to accurately assign a mass to each galaxy and then add up the total mass of all the galaxies in the cluster. This may be compared to the total mass needed to counterbalance the velocity dispersion in the cluster. This method is considered to be very reliable.

C. Effect of Gravity Versus Velocity Dispersion

For the galaxies studied in the Coma Cluster, the average speed of recession is about 7,000 kilometers per second. This is determined from studies of the red shifts of the light from these galaxies, which is considered to indicate a radial motion toward or away from the observer, assuming the red shift of the starlight is an actual Doppler effect. Each individual galaxy ordinarily has some smaller, random motion inside the cluster. This speed is around several hundred kilometers per second with reference to the neighbor galaxies.

The members should eventually escape from the Coma Cluster and wander off into intercluster space if there is not enough force to keep them in the cluster. If the universe is at least 4.5 billion years old, the random motions of the galaxies should have long ago disrupted the cluster and the galaxies could not possibly be as close together as they are now. As a matter of fact, there should be no cluster at all. The force that would counteract this escape tendency is the gravitational force of the mass of the cluster on the galaxy. The gravitational force of the matter in the earth pulls back a baseball thrown from its surface. In the Coma Cluster, the random motions of the galaxies would have to be balanced by the gravitational attraction of the matter in the cluster if the cluster is to stay together. This random motion of the galaxies in the cluster is called the velocity dispersion. Again, clusters of galaxies "appear" to be stable configurations. But since galaxies have random velocities in various directions with respect to the center of mass of the cluster, why haven't the galaxies escaped?

The velocity dispersion of the cluster can be calculated from the measured red shifts of the galaxies. The mass of a galaxy is related to its brightness. When the total mass of all the galaxies in the cluster is determined, the gravitational force can be calculated and considered with the observed velocity dispersion. In other words, we can calculate the amount of gravity that must be present to keep the galaxies together. Knowing the amount of gravity in turn allows us to calculate the amount of mass it would take to hold the clusters together. The result has surprised and astonished evolutionist astronomers. In the Coma Cluster the mass is too small to counterbalance the velocity dispersion by a factor of seven. In other words, for every 7 kilograms of mass necessary to hold the cluster together, only one kilogram can be accounted for. This is not a trivial matter. There is only fourteen percent of the matter in the cluster that should be there in order for the cluster to stay together. Astronomers have looked "high and low" for this "missing mass" but it is nowhere to be found. Things get worse in this search when clusters other than the Coma Cluster are studied; from two to ten times the needed mass is "missing" for many. For the Virgo Cluster, it turns out there should be fifty times more mass present than is observed. Ninety-eight percent of the mass expected is not found. This is called the "missing mass" problem. To sum up the situation: there is a vast discrepancy found when the mass derived from consideration of the motions of galaxies in clusters is compared with the mass that we can observe, based on the assumption that these clusters are billions of years old.

D. Attempted Explanations of the Missing Mass

Some have thought that the "missing mass" is located in intergalactic space. To be detected, the matter would have to emit some form of electro-magnetic radiation such as x-rays, visible light, or radio waves.

If cold matter exists between the galaxies, radio waves might be emitted and the radio astronomer could detect this. However, this has not been observed, and if small quantities of cold matter did escape detection, they would be far too small in amount to keep the clusters together.

A hot gas would emit x-rays. The background x-radiation that is incident on the earth's atmosphere can be explained by other means than the presence of a diffuse intergalactic material permeating space between the galaxies that is emitting x-rays. In the light of recent observational work, the matter emitting x-rays seems woefully small compared to what is needed. Certainly x-radiation has been observed associated with some galaxies, but the presence of the radiation has been explained rather well in terms not involving an intergalactic medium.

A slightly warm material would be difficult to detect since the radiation would be in the ultraviolet range of wavelengths which are mainly strained out by our atmosphere. However, using detection equipment in high-altitude rockets, balloons, and satellites, there has been no indication of a slightly warm gas between the galaxies.

The "missing mass" is not in the form of a diffuse gas in intergalactic space. Further conditions have been placed on this "missing mass." A study of the dynamics of the dispersion of the galaxies would indicate that the matter cannot be postulated as existing in one very massive object that does not have luminosity. The matter has to be distributed as a common constituent of intergalactic space. If someone says that alleged "black holes" are distributed between the galaxies, they would have to suppose these "black holes" to be as common as galaxies. As Bruce Margon[1] points out, there would have to be hundreds or thousands of them. There is no evidence for that. Again Margon[2] says the same objection applies to "dead" non-luminous galaxies or a large number of cool stars in between the galaxies of the clusters.

E. Conclusions

The obvious conclusion seems to be that the "missing mass" is not really missing since probably it was not there to start with. The universe, thus, could be quite young, and other lines of evidence strongly indicate this. The break-up time for these clusters (the time for dispersion of the galaxies so that there are no clusters) is far, far less than the alleged evolutionary age of the universe. This means that the clusters, since they have not been destroyed, are young, as well as the galaxies that form them. These galaxies contain stars that are alleged by the evolutionists to be the oldest objects in the universe (nine to twenty billion years old in the evolutionary scheme of things). This rapid break-up of the clusters coupled with their presence in the universe would indicate that these allegedly old stars are not old at all. It is believed by most astronomers that the Coma Cluster could not be younger than the Milky Way. So, if the cluster is young, the galaxy is young and the objects within the galaxy are young. The break up times of clusters are on the order of just a few millions of years at most. So the existence of clusters argues that the age of the universe has not reached anywhere near millions of years, which is much less than the age demanded by the evolutionists. The observations clearly indicate all galaxies are members of clusters.

It has been noted that the motions of the clusters look like those of bound systems which are not breaking up at all. If that is so, then the clusters would certainly be young, not having reached a stage where they are showing a looseness of organization indicative of a very old age.

To avoid the conclusions regarding time which are at the heart of evolutionary hypotheses, astronomers go to great lengths in attempting to invent explanations regarding the "missing mass." Margon[3] suggests that "we have reached an impasse, almost to the point Thomas Kuhn has called a scientific revolution. Apparently, unless the experimental data are blatantly in error, it is inevitable that some cherished astronomical or physical principle must fall." It would seem that the axe should fall upon the supposed aeons-long time age that is assigned a priori to the universe, the solar system, and the earth, since this concept of huge quantities of time leads to contradictory and illogical positions in certain aspects of astrophysics.

Chapter III
BRIDGES BETWEEN GALAXIES; SPIRAL GALAXIES; BREAK-UP OF STAR CLUSTERS

A. Bridges Between Galaxies

As mentioned previously, galaxies never occur singly. They are, found only in pairs or in larger aggregates. Observations appear to indicate that some pairs or multiples of galaxies are joined by bridges of luminous matter. In some cases the speeds of the galaxies along the radial direction alone are on the order of several thousand kilometers per second so that it would seem very unlikely that these galaxies are gravitationally bound. They seem, therefore, to have originated recently [by the creationist hypothesis as complete (or nearly completely) formed galaxies].

Some astronomers, such as Ambartsumian, have proposed that due to the apparent recent origin of these galaxies, and, consequently, their contents, they must have originated directly from explosions. In the light of the entropy law this sort of origin seems highly unlikely.

B. Destruction of the Arms of Spiral Galaxies

A galaxy is an assemblage of stars that does not rotate as a rigid body; the inner parts revolve more rapidly than the outer. An enormous difficulty which all hypotheses face that propose a long age for the universe is that any spiral arm structure will be almost completely wound up into a circle in one to a few (at most) revolutions of the galaxy—200 million to 1,000 million years— because of the Keplerian motion with which the parts of these galaxies rotate. However, there is still a huge number of spiral-arm galaxies in the universe. If they had a common origin in time, how could they still be around for supposed billions of years?

The magnetic field which runs through the gases in a spiral arm is not strong enough to give the arm appreciable coherence against the dragging effect of the rapid rotation of the nucleus. Further, the stars in the arms are not coupled to this magnetic field. In other words, the galaxy will wrap itself up in a short time, relatively speaking. This analysis does not, of course, determine the age of the universe; but certainly does establish limits which are far below evolutionists' estimates of the age of the universe, since there are very many spiral galaxies with the arms in well-defined existence.

C. Break-Up of Star Clusters

Many young, massive stars and also "old" (in the evolutionary scheme) stars are associated with groupings called "open star clusters." Dynamical studies show that open star clusters cannot last very long, because during the galaxy's ponderous rotation each star individually orbits around the nucleus. While the sun takes roughly two hundred million years to complete one revolution, stars closer to the nucleus require less time. Because each star moves in a separate orbit around the galactic nucleus, there is a shearing effect which tears star clusters apart. Star clusters are also subject to tidal distortions caused by the gravitational attraction of the galaxy's nucleus, and the individual stars are subject to perturbations caused by interaction with other nearby stars in the cluster. All these effects tend to disrupt star clusters, and calculations show that star clusters could rarely last more than a few hundred million to a billion years at the most. Some clusters can have been in existence for only ten thousand years at the most.[4]

Star clusters usually contain stars of all spectral types. Most astronomers believe that the stars and the clusters came into existence at the same time or had a common origin. There is no real evidence that stars are still forming today. The rapid disintegration of many star clusters would seem to argue strongly for a *young age* of the stars and the clusters.

Chapter IV

FORMATION OF INTERSTELLAR GRAINS

Cosmogonists have been puzzled mightily by the presence of interstellar grains. The density of available particles to form grains in interstellar space is very low—so low that the formation of grains appears impossible. If formation of interstellar grains (1×10^{-5} cm in diameter) on a basis involving naturalistic assumptions is nigh impossible, it would seem to be a tremendous "quantum" jump of presumption to suppose that star formation (an object ordinarily 1×10^{16} times larger than a grain) can be explained on a naturalistic basis. Also, we should consider the time involved in hypothetical grain formation as compared to the alleged age of the universe on an evolutionary basis.

A. Growth of Grains

To see the difficulties involved in this problem, consider the growth rate of a grain. I shall follow the approach of Martin Harwit.[5] Let us suppose to start with that there is a core of a grain existing at some time t with a radius which allegedly will change as time changes, or r = r (t). Suppose further that particles in interstellar space will impinge on this embryonic grain with a collisional speed v. Thus, it is proposed that the grain is moving through space with an effective speed v colliding with the atoms and molecules in interstellar space which presumably stick to it. If the number density of particles (number of particles in a unit volume of space) is n and each of these particles has a mass m, then the growth rate of a grain may be calculated rather simply

1. The grain's volume growth rate $= \left(4\pi r^2 \right) \dfrac{dr}{dt} =$ (surface area rate of growth) × (rate of change of radius).

2. The number of particles impinging on the embryonic grain per unit time $= n\pi r^2 v$ where n = number of particles/unit volume.

3. The mass of impinging particles per unit time $= nmv\pi r^2$ where m is the mass of a particle.

4. The added volume per unit time $= \frac{\pi r^2 nmv}{\rho} \alpha$ where ρ is the

the density of the interstellar atoms *after* they have been added to the grain and α is the sticking coefficient for atoms impinging on the grain.

Thus,

$$4\pi r^2 \frac{dr}{dt} = \frac{\pi r^2 nmV}{\rho} \alpha$$

and

$$\frac{dr}{dt} = \frac{nmV}{4\rho} \alpha$$

$$V \approx \left(\frac{3kT}{m}\right)^{1/2}$$

$$\frac{dr}{dt} \approx \frac{n(3mkT)^{1/2}}{4\rho} \alpha$$

B. Time for Grain Formation

We can make a few calculations to find the time involved in the alleged grain formation. Let

$$V \approx \sqrt{\frac{3kT}{m}} \qquad \text{with } T \approx 100° \text{ K}, \ m \approx 3 \times 10^{-23} \text{g}$$

$\rho \approx 3 g/CM^3$, $k = 1.380 \times 10^{-16} erg/k$, and $n \approx 10^{-3}$ particles / cm^3.

The maximum value of α is 1. Using these quantities, we get

$$\frac{dr}{dt} \approx n\sqrt{3kTm} \ \alpha/4\rho = 9.3 \times 10^{-23} \approx CM/s$$

The time for the grain to reach the size 10^{-5} cm would be

$$10^{-5} / 10^{-22} = 10^{17} s$$

This is approximately 3 billion years. Now with a more realistic value of $\alpha \lesssim 0.1$, the time for an interstellar grain to form would be $10^{-5} / 10^{-23} = 10^{18}$ sec or 30 billion years, which is larger than the evolutionist's estimate of the age of the galaxy. It should be noted that as the hydrogen is deposited on the grain, it would normally evaporate away very rapidly, thus, decreasing the radius.

Also, when the destructive effects are considered, it seems even more remote that interstellar grains could form. It is believed that in H II regions of the sky radiation pressure could often accelerate small grains to higher velocities than the larger grains. Inter-collisions would take place at such high velocities (v ≥ 1km/s) that both of the colliding grains would vaporize on impact. The vapors would have to recondense and the alleged process start all over. Sputtering by high-velocity protons can knock atoms off a grain's surface after they have become attached and reverse the alleged growth rate or, at least, slow the growth down.

C. Age Contradiction for the Universe from Grain Formation

From a consideration of the various effects involved, it would seem that grain formation is well nigh impossible. If it is granted that they could form, we are faced with the inconsistency for the evolutionist that the time of formation of the grains is greater than the alleged age of the universe. If it takes as long as indicated to form such a simple body as an interstellar grain as the calculations indicate under the most optimistic conditions (which actually do not seem to exist), how can the huge ages for the stars and galaxies postulated by the evolutionists have any credibility and thus be taken seriously?

Chapter V
OLBERS' PARADOX

Why is the sky dark at night? We certainly know that the night sky is basically dark, with light from stars and planets observed against a dark background. But a bit of thought will show that if there is a uniform distribution of stars in space, it should not be dark anywhere in the sky. If we look in any direction at all we will eventually see a star, so the sky should appear uniformly bright. We can make the analogy to our standing in a forest. There we sh.. ... .se some trees that are closer to us and some trees that are farther away, but if the forest is big enough our line of sight will always eventually stop at the surface of a tree. If all the trees were painted white, we would see a white expanse all around us. Similarly, when looking up at the night sky we would expect the sky to have the uniform brightness of the surface of a star. The fact that the sky is dark at night is called Olbers' Paradox. Olbers phrased his paradox in terms of stars, whereas the stars are actually grouped into galaxies. The same argument applies to galaxies and we must see the average surface brightness of galaxies everywhere. This paradox is obviously involved in the question of the age of the universe.

A. The Paradox Phrased Mathematically

Suppose that space were filled with stars. The light emitted by the stars in a shell at a distance r to r+dr from an observer would be proportional to $(4\pi r^2)$ (dr) = (surface area of the shell) × (thickness of the shell). Of this light, a fraction, proportional to $1/r^2$, would be incident on the observer's telescope, since light intensity drops as the inverse square of the distance. Thus, from each spherical shell of thickness dr an observer would receive an amount of light proportional to dr alone since, amount of light $\propto 4\pi r^2 \, 1/r2$ dr $\propto$ dr. On integrating this expression out to infinite distance the light received by the observer should have an infinite value:

$$\text{Total Amount of light} = k \int_0^\infty dr \to \infty \; .$$

This infinity arises only because selfshadowing of stars has not been taken into account. A foreground star will prevent an observer from seeing a star in a more distant shell, provided both stars lie along the same line of sight. When shadowing is taken into account, we find that the sky should only be as bright as the surface of a typical star, not infinitely bright. That is still much brighter than the daytime sky; the night sky is still much fainter.

If one believes in a very large size for the universe and a huge age for the universe this situation would appear paradoxical. Olbers first mentioned this in 1826.

B. Attempts to Resolve the Paradox

1. The introduction of "curved space" (whatever the relativists mean by that) cannot circumvent the paradox. The surface area of a sphere is of the form $S(x) = 4\pi a^2 \sigma^2(x)$ and is a function of distance x alone. The number of stars in a spherical shell is proportional to $S(x) dx$. But the amount of light reaching the observer from that shell is also reduced by a factor $S(x)$, and these two factors cancel to give the same distance independence obtained for a flat space.

2. It has been argued that interstellar dust might absorb the light. But in a very old universe, dust would come into radiative equilibrium with stars and would emit as much light as was absorbed. The dust would then either emit as brightly as the stars, or else it would evaporate into a gas that either transmitted light or emitted as brightly as the stars.

3. Unless one wishes to say no laws of physics hold for phenomena on such a scale, we are led to one of the three conclusions:

a) *The density or luminosity of stars at large distances diminishes.*

b) *The constants of physics vary with time.*

c) *There are large systematic motions of stars that give rise to spectral shifts.*

Argument a) would hold if the universe were very young—stars would have only been radiating a short period of time.

Argument b) has no evidence whatsoever for it.

Argument c) states that the expanding universe need not be bright since the radiation from distant galaxies is less intense by the time it reaches the observer. Photons reaching the observer from points close to the cosmic horizon, where the red shift of galaxies approach infinity, will have an energy and arrival rate approaching zero. However, the expansion of the universe has been challenged on a number of counts. If the red shifts do not all represent a real expansion of the universe, then this argument is not an answer at all or at best only a partial answer.

C. Conclusions About Time from the Paradox

The universe having a dark night sky seems in the consideration of other arguments persuasive evidence of a young universe. The presence of very bright objects seems indicative of a low entropy state and a short time that the universe has been running down since its beginning. These bright objects (and there is an infinite number of them, practically speaking) have not been in existence long enough to give a bright night sky.

Chapter VI

TRAVELLING OF LIGHT IN SPACE

Since the solar system and Earth are a part of the universe and thus must be intimately related to the stellar systems, it is appropriate to discuss an objection that is raised whenever it is suggested that the age of the universe is quite young. The objection is as follows: in spite of the physical indicators that seem to show that some of the celestial objects are young, these indicators must be somehow basically wrong if applied to the universe as a whole because "it is known that some of the stars are literally billions of light years from us." We know that light travels at the rate of 186,000 miles per second in a vacuum; thus, if a star is at, say a distance of a billion light years from us, it would take a billion years for the light from that star to reach us. Therefore, it is pointed out that since these objects are very far from us, the light from these objects requires huge spans of time to reach Earth. The universe, then, would have to be as old as the length of time for the light to travel from the very distant stars and galaxies.

A. Distances of Stellar Objects

There are various aspects that could be discussed regarding this problem of distances to the stars since it is a far more involved and difficult problem than it is made to appear in many astronomy textbooks and some astronomical papers.

Perhaps it would be worth our while to take a brief look at the methods for finding distances in the universe. The universe is mapped by a series of techniques, each of which moves us out to a greater range of distances—to the next level of the celestial distance hierarchy. It would seem that each level of the hierarchy is less reliable than the previous one, so that there is very much uncertainty about measurements of very great distances.

Let us consider the so-called "parable of the city" to show the methods involved. Suppose a Venusian (if there were such a being) lands in the dark of night on a flat roof of a building in El Paso. "He" wishes to map the city, but is confined to the roof from which he can see the faint outlines of objects. Outside, only lights are visible: street lights, traffic, late-burning lights in the rooms of hard-working (or otherwise) students, etc. Distances between objects on the roof can be found by direct comparison with any convenient standard, such as the Venusian's "foot." Distances of lights in the immediate neighborhood can be found by triangulation using a baseline joining any two points on the roof. But the size of the roof is limited, so that to map more distant lights he must devise another method. In fact, he employs the decrease in their apparent luminosity with distance: the assumption is made that lights with the same spectral characteristics (such as the red-yellow-green of traffic lights) are physically identical, so that the inverse square law can be used to determine distances. Provided there are "standard candles" available for calibration in the triangulable "near zone," this method works out to distances at which the lights are so faint as to be barely visible. Farther away, the only visible objects are accumulations of lights—apartments, suburbs—whose distances can again be determined from luminosity measurements, provided the nearest such objects contain visible traffic lights, etc., whose distances are known. Thus, the Venusian maps the city by using a heirarchy of methods of overlapping applicability, and thus transcends the limitations imposed by his restricted viewing point. Let us take a look at how this works in the universe.

1. Parallax

This is triangulation using a baseline AB whose length 2d is known (see Fig. 1), oriented so that the perpendicular from O, the object whose distance D is to be measured, joins AB at its midpoint. As we move from A to B, the direction of O will change. This can be observed as an apparent displacement of O relative to objects far more distant (whose direction hardly alters). If the angular displacement of O is 2θ, then D follows by elementary trigonometry as

$$D = d\cot\theta \approx d/\theta$$

since in practical cases θ is a very small angle. The angle θ is called the parallax of O relative to AB.

The smallest parallax that can be measured as determined by the

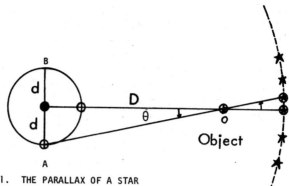

Figure 1.   THE PARALLAX OF A STAR

By definition, a star whose parallax is 1
second of arc is at a distance of one parsec.
If it has a parallax of $\frac{1}{2}''$, or 0.5'', it
is at a distance of two parsecs. In
general,

$$d(\text{in parsecs}) = \frac{1}{\theta''} .$$

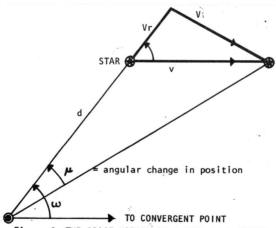

Figure 2. THE SPACE MOTION OF A STAR IN A MOVING CLUSTER.
Omega is the angle between the star and the
convergent point; v_r is the radial velocity;
v_t is the tangential velocity and v is the space
motion.

resolving power of the largest telescope is 0,05", and this limits us to stars closer than about 20pc or 66 light-years. The parsec (pc) is the distance out to a hypothetical star whose parallax is 1".

2. Distance from Velocity Measurements

The "fixed stars" move, with velocities $\vec{V}$ which can be resolved into two components: a radial velocity V_r along the line of sight (Fig. 2) and a transverse velocity V_t perpendicular to the line of sight. We can measure V_r by the Doppler shift $\Delta\lambda$ of a spectral line of "rest wavelength" in the light from the star. λ is identified by comparison with patterns of lines observed in terrestrial laboratories. Then,

$$V_r = \frac{c\,\Delta\lambda}{\lambda}$$

a positive V_r (recession) is indicated by a shift of the lines towards the red side of the spectrum, and a negative V_r (approach) is indicated by a blue shift. It is not possible to measure V_t directly, but for nearer stars we can observe the angular velocity which is due to V_t; so, $D\omega = V_t$ where D is the unknown distance of the star, and ω is called the proper motion of the star.

D cannot be determined simply by measuring V_r and ω, but there are two limiting cases where extra information is available. In the first, we use a cluster of stars which all have approximately the same velocity $\vec{V}$. This parallel motion can be recognized by the convergence of the proper motions on a point (direction) in the sky (see Fig. 2). The angle ϕ between this direction and the line of sight to the cluster is the angle between V_r and V, and we can write

tan $\phi = V_T / V_r = D\omega / V_r$.

Therefore, $D = V_r \tan\phi / \omega$

This moving cluster method works best for open clusters. For the Hyades cluster in Taurus this will put us out 40.8 pc or about 135 light years. The unfortunate aspect of this is that there are few clusters for which a convergence point can be determined.

The second limiting case uses groups of stars with no overall motion. Where the stars appear to have random velocities like the molecules in a gas, it is assumed that the random motion is "isotropic," that is,

$$(2V_r{}^2)_{av} = (V_T{}^2)_{av} = D^2(\omega^2)_{av}$$

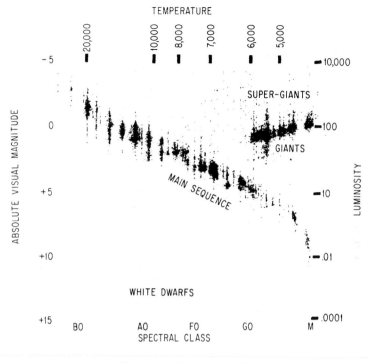

Figure 3.   THE H-R DIAGRAM.

and

$$D = \sqrt{(2V_r{}^2)_{av} / (\omega^2)_{av}}$$

involving averages over the measured V_r and ω for a sample of stars in the cluster. This method of statistical parallaxes takes us out to several hundred parsecs ≈ few thousand light-years,

3. Distance from Apparent Luminosity

Suppose we know the absolute luminosity L of a star or galaxy; this is defined as the total power radiated. Suppose also that we measure the apparent luminosity ℓ; this is defined as the power traveling normally across unit area at the observer. Then, if there has been no absorption in space we may use the conservation of energy to derive the inverse-square law for ℓ in terms of L and the distance D:

$$L = (\ell)(4\pi D^2)$$

therefore,

$$D = \sqrt{L/4\pi l}$$

Thus, D can be found from measurements of l, provided L is known; this L is generally not known. However, there are certain classes of objects for which it is "believed" that L is known, usually not very accurately, and all higher levels of the cosmic distance hierarchy are based on these "standard candles."

Astronomers do not use the l as a measure of apparent brightness but use a logarithmic measure, the apparent magnitude m. This m is greater the fainter the object and is defined so that two objects whose ratio of luminosities $l_1/l_2 = 100$ differ in apparent magnitude by 5, that is,

$$l_1/l_2 = 100^{(m_2-m_1)/5} ; \quad m_2 - m_1 = 2.5 \log_{10} (l_1/l_2).$$

The absolute magnitude M of an object is the apparent magnitude that the object would have at a distance of 10 parsecs. Thus, measuring D in parsecs,

$$l_m \times 4\pi D^2 = l_M \times 4\pi 10^2.$$

Therefore,

$$100^{(M-m)/5} = 10^2/D^2 ;$$

$$D = 10^{[1+ (m-M)/5]} ;$$

$$m - M = 5 \log_{10} (D/10)$$

It is believed that M for several classes of objects (main-sequence stars, Cepheid variables, novae, and brightest galaxies in clusters) can be determined on the basis of certain assumptions:

a) Main sequence stars

For nearby stars whose distances can be found by parallax or velocity measurements, Hertzsprung and Russell found in 1910 that for many stars (the "main sequence") L (or M) and the spectral type (roughly the color, which corresponds to the surface temperature T) are strongly correlated (Fig. 3); the cooler stars are fainter. Thus, if we know that a given star is in the main sequence, we simply measure its apparent magnitude m and determine its spectral type; the latter may then be used to give us the absolute magnitude M from the Hertzsprung-Russell diagram. From the distance modulus, m–M, we infer D. The method works best for clusters where all the stars are believed to be roughly at the same distance, so that the main sequence can be identified in a statistical way, as follows: the apparent magnitudes m are plotted against spectral type, and it is often found that the points lie near a curve similar to that resulting from the Hertzsprung-Russell relation; from this it is assumed that any star on the curve is a member of the main sequence. Complications arise bacause there are (roughly speaking) two main sequences: population I stars (like the Sun) in open clusters and population II stars in globular clusters. Main-sequence stars are rather faint (at least where the color-luminosity relation is considered reliable) and this limits the distances which can be determined. The telescope cannot detect stars fainter than about m = 22.7. If such a star is seen to have the same color as the Sun (M = 4.7), then the distance modulus is m–M = 18, and the distance is

$$D = 10^{1+18/5} \ pc \approx 4 \times 10^4 \ pc \approx 130,000 \text{ light years.}$$

b) Cepheid Variables

Many stars vary regularly in brightness, with periods (P) on the order of days. Typical is δ Cephei. It was noted by Henrietta Leavitt in 1912 that m and P are approximately linearly related for the cepheids in the Small Magellanic Cloud. Since all the stars in the cloud are assumed to be at roughly the same distance from us, she concluded that since m was uniquely related to M, and for the cepheids a direct m to P, an absolute luminosity versus period relation may be deduced. Because all Cepheids with the same period are assumed to have the same M, these stars may be employed as "standard candles" for distance determination, once the relation has been calibrated by establishing M for Cepheids within the galaxy, using a lower level of the distance heirarchy. Unfortunately, there are only a few galactic cepheids in clusters whose distances are known, and this reduces the accuracy of the method.

Now Cepheids are intrinsically rather bright and they can be resolved in some external galaxies as well as our own. Cepheids can be seen out to about 4×10^6 pc $\approx$ 1,300,000 light-years.

c) Novae

In each large galaxy close enough to be studied in detail, about forty stars are observed each year to flare up suddenly and become up to six magnitudes brighter than they were before. Then they gradually fade over a period of days. These are novae, or "new" stars. They can also be observed in our own galaxy, and their distances determined by methods lower down in the hierarchy. Thus, the absolute magnitudes M are known. These vary with time and it is believed that the maximum brightness M_{max} is fairly well correlated with the fading time. Thus novae can be used like Cepheids to infer D from the fading time and the apparent maximum brightness m_{max}. This puts us out to a distance of 4×10^7 pc $\approx 13,000,000$ light-years. We have now reached out beyond the nearest galaxies outside our local group; these lie in the Virgo Cluster, which contains about 2,500 galaxies. Extra checks on the estimation of distances on this level of the hierarchy are obtained by using globular clusters of stars and HII regions (clouds of ionized hydrogen surrounding hot stars) as "standard candles."

d) Brightest Galaxies in Clusters

Beyond the Virgo Cluster it is not easy to resolve individual stars, and the final level of the distance hierarchy employs whole galaxies as distance indicators. It appears that the distribution of apparent galactic brightnesses within a cluster has a rather sharp upper limit. For the Virgo Cluster we know the distance and, hence, the absolute magnitude M of the brightest galaxy. Its value is –21.7. If we assume the brightest galaxy in a distant cluster has the same M, then we can find D simply by measuring the apparent magnitude m. This moves us out to

$$D = 10^{1 + [22.7 - (21.7)]/5} \, pc \approx 8 \times 10^9 \, pc \approx$$

26,000,000,000 light-years.

At this point it is important to draw attention to a problem in this regard that will be important later in our discussion. The problem concerns the simple distance formula given by the parallax, velocity, and luminosity methods. The formulae involve elementary trigonometric relations between angles and distances for systems of straight lines. These are based on ordinary Euclidean geometry, together with the identification of the light rays with straight lines. This is known to apply very accurately in the solar system because the various very precise methods for determining distance agree with one another. But can this be extrapolated on out to greater distances? There are various systems of non-Euclidean geometry which might be applicable. It is a question not of mathematics but of physics to ask which geometry applies to the trajectories along which light travels.

B. Travel Mode of Light

In connection with the time it takes for light to get here from the stars, a very important work was done by Parry Moon of the Massachusetts Institute of Technology and Domina E. Spencer of the University of Connecticut. This work was published in the *Journal of the Optical Society of America,* August, 1953. This work is significant particularly in the matter of the age of the universe as well as it is to cosmology and to electrodynamics.

This paper involved a re-examination of the alleged proof of one of Einstein's basic assumptions in his hypothesis of special relativity. Einstein made the very basic assumption that the speed of light which an observer measures is independent of the motion of the observer, and is also independent of the motion of the source of light. An observer in any frame of reference in space would measure roughly 186,000 miles per second for the speed of light in a vacuum.

This idea was put forth by Einstein as an assumption and it was very basic to his ideas on relativity. It is an idea that is actually contrary to all human experience, and it certainly does away with common sense ideas of space and time. The Swiss physicist, Walther Ritz, about the same time that Einstein was formulating his relativistic concepts, proposed an opposite view. Ritz said the speed of light is dependent on the speed of the object radiating that light. He said the speed that an observer would measure does depend on the speed of the star that is radiating that light as well as the observer's speed. His theory was called the "emission theory," or as some people today call it, the "ballistic theory." Ritz said that when the speed of light was calculated, a factor for the speed of the star should be put in the consideration. The Dutch astronomer de Sitter proposed a test for the opposing ideas of these two men.

Using Ritz's ideas, de Sitter set up a mathematical equation dealing with a binary star system. A binary star consists of two stars that travel about a common center of gravity in space. The stars travel in separate orbits about this common center of gravity. The center of gravity between the two stars is sort of a balance point. For example, if you should put a rod between the stars connecting them, the balance point of the rod (like a fulcrum of a seesaw, for instance) would be called the center of gravity of the two stars.

Taking Ritz's statements, a characteristic constant was formulated by de Sitter that would describe the appearance of the double-star system:

$$\Gamma = (2\pi/p)\,(r/c)\,\left(\frac{v}{c}\right)$$

where v is the speed of the star in its orbit about the more massive star; p is the period of the star; r is the Euclidean distance of the center of gravity of the star system from the observer in the plane of the orbit of the star; and c is the speed of light in a vacuum. The period is the length of time it takes the star to swing around completely in its orbit. If the apparent motion of a star as seen by a distant observer is plotted, interesting results are obtained. For $\Gamma = 0$, the motion is a true sinusoid. As Γ is increased the curve is tilted and the function may become multivalued. This means that more than one image of the star may be seen. If $\Gamma > 1$ there should be more than one image of that binary star. In other words, you should have multiple images of that star. For $\Gamma < 1$ no multiple images should occur. If you are using a spectroscope and recording the spectral lines of the light from the star, for $\Gamma > 1$ there will be multiples of those lines on the spectral plate.

Twenty-seven visual binary stars were studied by de Sitter. The Γ's for those stars were calculated. The Γ's for all these visual binaries were all much less than 1. Now what does that mean? You have no test at all if Γ comes out less than 1, because according to de Sitter's interpretation of Ritz's hypothesis there would be no multiple images. There would only be multiple images of those stars for which Γ turned out to be greater than 1. Contrary to many statements, the Ritz theory does not predict multiple images or multiple spectral lines for the visual binaries de Sitter used.

Rarely is it mentioned how the Γ's had turned out in this case. It is just stated that there are no multiple images for those stars, and, thus, Einstein was right. Seldom does anyone bother to note that the inconclusiveness of the data was similar to that in the Michelson-Morely experiment which did not show there was an ether; but it did not show there was not an ether either. It remained an inconclusive matter so far as the results of the experiment. But many physicists, of course, have used the Michelson-Morely experiment as evidence against the idea of an ether pervading space between the stars and the planets. However, today physicists are talking about some kind of ether again. Not the old eighteenth or nineteenth century ether, but, an ether of some sort or other pervading space.

Moon and Spencer decided to go a bit further in this matter. There have been a lot of binary star systems discovered since de Sitter's days. They took spectroscopic binaries, and calculated the Γ's. They considered some cepheid variables which are actually pulsating variable stars. These are stars that oscillate like the beat of a heart. They considered a few eclipsing binaries as well. They calculated the Γ's for them. Many of the Γ's turned out to be greater than 1. This meant that they should see multiple images, according to Ritz's idea. However, no multiple images have been observed and they concluded that either the Ritz hypothesis must be dropped, or a radical change must be made in the distances of velocities of the stars. Moon and Spencer believed Ritz was correct since common experience here on Earth shows that the speed of the source does have something to do with the speed of the wave emitted by the source.

The major possibilities are that Einstein was right in saying that the velocity is constant with respect to the observer; that the velocity of light is constant with respect to the source only (Ritz's hypothesis) but the Γ's need to be changed in some manner by either reducing the distances of the stars or changing the period of the binary system, or assuming the Doppler shift is not a true velocity effect.

Moon and Spencer considered that if we take the Ritz's hypothesis as correct, there is a way to change the Γ's so that they will turn out less than one. If light, instead of travelling along straightline paths (or what are called Euclidean paths), travels along curved paths on curved surfaces, then the distance, which is the factor causing the Γ to be greater than 1, will be changed.

The German mathematician, Georg Friedrich Bernhard Riemann, had proposed a curved space geometry long before this matter arose. There are different kinds of curved space geometries, but Riemannian geometry is what Moon and Spencer have suggested as the answer to the Γ's turning out greater than 1. This gives the possibility that Ritz was correct in what he said.

Moon and Spencer proposed a Riemannian surface on which light would travel in curved paths. They took the data regarding binary stars and calculated a radius of curvature for the proposed Riemannian surfaces. They calculated a curvature for the Riemann surfaces that would best represent the data on binary stars to give Γ's less than 1. Now, in the equation for γ, if you change from Euclidean distances over which light travels to Riemannian distances and change to a velocity for the star calculated on the basis of Riemannian geometry, a most interesting result is obtained:

$$\Gamma_E = \left(\frac{2\pi}{r_c^2}\right)(r)\left(\frac{dr}{dt}\right) \quad \text{(Euclidean)}$$

$$\Gamma_R = \left(\frac{2}{r_c^2}\right)(s)\left(\frac{ds}{dt}\right) \quad \text{(Riemannian)}$$

Where $\frac{dr}{dt} = v, \frac{ds}{dt}$

is the corresponding velocity on a Riemannian surface and s is the Riemannian distance corresponding to r, the Euclidean distance. The conversion metric between the two geometries is $s = 2R \tan^{-1}(r/2R)$ where R is the radius of curvature of the Riemannian surfaces. In the solar system Euclidean and Riemannian distances are practically the same since the distance is too small a distance to give much of a difference between Riemannian and Euclidean geometry. The distance out to Pluto begins to show just a slight difference, however, between the Riemannian and Euclidean distances.

There is not much change between the two until the nearest star, Alpha Centauri, is reached. Consider the distance on a Riemannian surface of a star out on the edge of the observable universe at an infinite distance in Euclidean geometry. Upon taking the Euclidean distance as infinite, and a radius of curvature of five light years for the Riemannian surface, the distance to that star in Riemannian geometry according to the metric conversion is only 15.71 light years. That means if light travels on a Riemannian surface, it would take only 15.71 years of sidereal time to get from the farthest star to us. Table I gives you an idea of the conversion between the two geometries.

TABLE I	
(After Moon and Spencer)	
Euclidean Distance	**Riemannian Distance**
(light years)	(light years)
1	0.997
4	3.81
30	12.5
100	14.7
1000	15.6
10000	15.7
infinite	15.71

C. Age of the Universe and the Path of Light

If light is traveling on Riemannian surfaces with these Riemannian distances, then the argument that the universe cannot be young since it takes light so long to get here, is not a valid position. You can leave the stars at their astronomical locations, in Euclidean space, but the light from these stars can get to us in very small periods of time—at the most 15.71 years.

The Riemannian distance goes toward a fixed upper value while the Euclidean distance goes to infinity. The nature of the conversion metric $s = 2R \tan^{-1}(r/2R)$ between the two distances has in it the factor $[\tan^{-1}(r/2R)]$ which prevents the Riemannian distance from becoming infinite. It is still to be determined what is correct in this matter. But it would seem that there is the possibility of having a huge universe but one quite young in age.

It should be remarked that there are a number of other suggested solutions to this problem that also give a young universe, though retaining its present alleged size.

Chapter VII
COSMIC DUST INFLUX

A. Cosmic Dust Influx of Earth

One of physical indicators of a young age of the Earth-Moon system has to do with the influx of micrometeoric dust into the earth's atmosphere from interplanetary space and finally down onto the earth's surface. Eventually, some of the dust is carried off the continents by river action into the oceans and the deep-sea sediments. This dust material is called micrometeoric since the dust particles are very small, being only a few ten thousandths of a centimeter in diameter at the maximum size and somewhat smaller at the minimum. These particles (popularly called cosmic dust) do not burn up with entry into the atmosphere since their ratio of surface area to mass is so large. They settle gradually to the earth's surface. If there are constituents in the cosmic dust that are not very common on the earth, the influx of these elements may serve as a "clock" since these constituents would furnish the main supply of those elements to the earth.

One method for finding the influx rate of cosmic dust involves the collecting in chemical trays of matter settling down through the atmosphere to the earth's surface, and then the analysis of this material to find what is truly extraterrestrial in it. Only the matter that is magnetic, and of this only the particles that have shapes suggestive of having been aerodynamically affected by the atmosphere are used in the calculation since stony meteoric matter cannot always be clearly separated from terrestrial matter in this approach. Consequently, estimates of the influx of meteoric dust on this basis are very conservative since it is now considered that stony meteorites are far more abundant than iron meteorites. Over 90% of all meteorites are believed to be stony in composition.

Another method involves the study of satellite impacts with these micrometeorites in interplanetary space. From a study of the signal generated by the impact, the size distribution of these particles and the mass of these particles can be calculated. This gives a second and probably more reliable determination than the chemical-tray-collector approach. There can be no terrestrial contamination of the cosmic dust sampled by this method.

Estimates of the influx of cosmic dust on Earth's surface range considerably, with different investigators[6] (from 10,000 tons/day to 700,000 tons/day). The Swedish geophysicist, Hans Petterson,[7] estimates 14,300,000 tons of meteoric dust come onto the surface of the earth per year. Consequently, if Petterson's figure is used, in five billion years there should be a layer of dust approximately 54 feet to more than a 100 feet in thickness on the earth, depending on the density of matter, if it were to lie undisturbed without the various erosional agents acting on it. Using the 700,000 tons/day value, the layer should be about 965 ft. in thickness. Because erosional agents are acting, some of the extraterrestrial nickel (since it is one of the major constituents of meteorites) will be carried into the oceans. There it will add to the amount going into the oceans directly from the settling through to the atmosphere.

B. Age of the Earth from Oceanic Nickel Content

Nickel is not very abundant in terrestrial rocks and the continental sediments and is, relatively speaking, uncommon in ocean water and the oceanic sediments. This seems to indicate a rather short age for oceans. Taking the amount of nickel in the ocean water and ocean sediments and using the rate at which nickel is being added to the water from meteoric material, the length of time for accumulation of the nickel turns out to be only several thousand years, rather than a few billion years.

Looking at the problem purely on the basis of runoff water from the continental masses to the oceans, a most interesting result is obtained. The computed mass of the ocean waters is 1.417×10^{21} kg., and runoff from the continents (as river water) contributes 3.333×10^{16} kg. annually. At the present rate of influx, assuming no withdrawal of water by evaporation or other means, this would fill the ocean basins to their present volume in only 44,500 years—a period called by geologists "the residence time" of ocean water. Given the respective concentrations of an element in river water (X_r) and in the ocean (X_o), the residence time (t_o) for an element can be calculated by the formula:

$$t_o = \frac{X_o}{X_r} \; (0.0445 \times 10^6) \; years$$

The nickel concentration in ocean water is 0.002×10^{-6} gm/gm of ocean water and in river water is 0.01×10^{-6} gm/gm of river water. The residence time for nickel, then, is 8,900 years. The nickel in the oceans can be accounted for from continental runoff alone in only 8,900 years. That value will be reduced somewhat when taking into account the influx of cosmic dust from the atmosphere.

From core samples of deep-sea sediments, there is a near absence of the cosmic spherules (cosmic dust that has been somewhat shaped because of travel through the resisting atmosphere) which come from interplanetary space, but quite an abundance, relatively speaking, of meteorite ablation products (material produced in the destruction of a meteor as it travels through the air). This would seem to be a further indication that the ocean sediments are quite young.

C. Dust on the Moon

From the reports of the first lunar landing, the accumulation of dust on the surface of the moon in the vicinity of the touch-down was very small (a layer not much more than 1/8 inch in thickness). The later landings were in "seas" that have larger dust accumulations, but these layers were still very small in thickness compared to the earlier predicted amount.[8] The dust layer is ⅛" to 3" in thickness. The moon moves through the same region of space that the earth does and, consequently, should have about the same influx of cosmic dust on its surface as on the earth. Astronomers had been concerned that a lunar spaceship upon landing would sink into the supposed huge amount of dust that should have accumulated on the surface of the moon in about 4.5 billion years of assumed time.[9,10] The rocket would be stuck in this layer of "mud" and not be able to leave the moon. Also, in the "sea" areas, where lunar ships landed, there should have accumulated more dust than elsewhere on the moon. Yet, the amount of dust is amazingly small. What could have happened to all the dust, assuming a 4.5-billion-year-old moon?

Another contributing source to the dust layer on the moon has been suggested by R. A. Lyttleton[11] of Cambridge University. Since there is no atmosphere on the moon, the moon's surface is exposed to direct radiation. Thus, the strong ultraviolet light and x-rays can destroy the surface layers of exposed rocks and reduce them to dust at a rate of a few ten-thousandths of an inch per year. If a layer, say 0.0004 inch thick of pulverized matter, is formed per year, then, in 10,000 years a layer about four inches in depth would be produced; in 100,000 years a layer of 40 inches; in 1,000,000 years a layer of 3.3 feet; in 1,000,000,000 years a layer of 6.3 miles; and in 4,500,000,000 years a layer about 28 miles in depth would be formed. Thomas Gold speculated that electric fields have swept the surface of the moon carrying dust material into the seas so that certainly many miles of this material should have accumulated if the moon is 4,500,000,000 years old. But the dust and debris layer is not there to any appreciable depth.

From the Apollo missions it is known that the bedrock of the lunar surface is covered by a thin, well-mixed layer of loose soil and rocks —the regolith. The regolith is believed to be produced by meteorite impacts which fragmented the bedrock, and also added to the pulverization that should have taken place as mentioned above from high-energy radiation. The regolith is estimated to have a depth of about 18 feet at maximum in the maria and about 4 inches depth on the highlands. Again, upon considering an additional effect of meteorite impacts pulverizing and changing the surface, no depth is obtained approaching that required for a time scale on the order of 4.5 billion years.

Although more data and calculations are needed to make a quantitative study of the age of the moon, the order of magnitude of its age on this basis seems to demand no more than just several thousand years.

Chapter VIII
COMETARY LIFETIMES

A. Physical Nature of Comets

Comets have come to be considered as perhaps the most puzzling and mysterious of all the many types of celestial bodies. Their different properties are nearly the antithesis of those of the planets, and comets are sometimes referred to as "the second solar family"—a kind of extra solar system. The word "comet" comes from the Greek word *komé*, or hair, since a comet resembles a "long-haired star." Orientals knew comets by the descriptive name "broom stars." The name thus fancifully refers to the most obvious features of many of the comets which actually do resemble a star embedded in a mildly luminous fog from which there appears a long streaming trail, itself faintly luminous.

All comets move in elliptical orbits about the sun and are thus members of the solar system. There are two distinct classes of comets: (1) the long-period comets ($P = 10^2$ to 10^6 years) which are in the great majority and have orbital eccentricities very close to 1 (almost parabolic or hyperbolic orbits) and (2) the much smaller group of short-period comets ($\bar{P} = 7$ years) which have orbits with eccentricities considerably less than 1. Most comets reach perihelion around one or two astronomical units (approximately 93 to 186 million miles) from the Sun, and the inclinations of their orbits range through all values. Comets move in a volume of space centered on the sun while the planets move near the ecliptic plane. The aphelia of long-period comets lie in the range roughly from 10^4 to 10^5 AU, so that their orbital periods are vast. The observed long-period comets have only been seen during the course of one perihelion passage. Consequently, the data are quite scanty regarding long-period comets. In some cases, planetary perturbations transform some comet orbits from ellipses to hyperbolas ($e > 1$)

so that they then escape from the solar system never to return. Of the short-period comets, Comet Encke has the shortest period (3.3 years) while Comet Rigollet has the longest (151 years). One of the most famous short-period comets is Halley's comet which returns approximately every 76 years (the next appearance being around 1986).

Current models of comets are based upon their appearance and their spectra. Far from the sun we see only a star-like object which, it seems, shines by reflected sunlight. Since these nuclei do not noticeably perturb the orbits of planets or their satellites, their masses are 10^{-6} to 10^{-8} the earth's mass. Since they are not seen when passing across the face of the sun, their radii are probably smaller than 50 km. The density of the nucleus is around 1 gm/cm^3. Such a nucleus has been described as a tight swarm of ice and dust particles (a "flying gravel bank") according to R. A. Lyttleton, or as the dirty iceberg (dust and stones bound in a matrix of frozen water, carbon dioxide, methane, and ammonia) proposed by Fred Whipple. When the comet is a few AU from the sun, a tenuous gas halo or coma about 10^5 km across is formed. The nucleus-coma combination is called the head of the comet. Closer to the sun at I-AU, when the comet can be seen in the twilight sky of sunset, a tail or several tails are seen to form in some comets. The tails always point away from the sun, and they are of two types: (1) a curved dust tail believed to be formed of solid particles escaping the nucleus into their own solar orbits, and (2) a straight gas tail pushed back from the comet's head by solar radiation and the solar wind. Occasionally, explosive action in the nucleus produces secondary tails.

B. Destruction of Comets

With each perihelion passage it is observed that short-period comets lose gas, dust, and rock as a result of the intense solar heating of the comet, the tidal forces exerted by the sun and the large planets, and the solar wind so that eventually the comet vanishes. Some short-period comets have been observed to split into several pieces and even disintegrate. Certain comets, known as the sun-grazers, have passed within about 0.005 to 0.007 AU of the sun's surface at perihelion. These sun-grazers all have similar orbital elements. It is thought that these comets may all be pieces of a single-parent comet that broke up after a close encounter with the sun. The sun-grazing comet 1882 II broke into several pieces during perihelion passage. First discovered in 1772, Comet Biela (with a period of 6.6 years) was seen to break in two in 1846 and then was observed as a double comet in 1852, but disappeared from observation thereafter. In 1872 a brilliant meteor shower was seen as the earth crossed Biela's original orbit; the cometary debris strewn along the orbit had entered the earth's atmosphere as meteors. The Biela shower now occurs on November 14 of each year as the Andromedids. In 1965 Comet Ikeya-Seki emerged from the solar corona as two pieces that separated at 25 mph. This bright comet was visible even during the daytime. Finally, several comets have disappeared entirely near their perihelia: Comet Ensor 1929 III is an example. Detachment of some portion of a comet has been observed on ten occasions for different comets during the present century.

C. Lifetime of Short-Period Comets

The lifetimes of the short-period comets have been estimated variously by different investigators. By lifetime is meant the time elapsed from the origin of the comet to its destruction. The estimates have ranged from R.A. Lyttleton's calculation[12] of ten thousand years to Fred Whipple's calculation[13] that on the average a short-period comet will make two hundred trips around the sun during its lifetime. Thus, from Whipple's estimate, 1,400 years would be the lifetime of an average short-period comet since the average period of these comets is seven years. Obviously these estimates are far apart, but several thousand years seems to be the best estimate of the lifetime of a short-period comet. *This seems then to put an upper limit on the age of the solar system,* if the comets came into existence at the same time as the solar system. Thus, the solar system is quite young.

It is objected that this may not be so for the supply of comets may be replenished as time goes by through different mechanisms, thus nullifying this conclusion. I would like to examine the different ideas regarding the origin or source of short-period comets suggested by different investigators and consider the validity of their hypotheses.

D. Discussion of Hypotheses Regarding Cometary Origin and/or Source

There have been a number of hypotheses proposed to explain away the obvious young age of the comets and the consequent young age of the Solar System. Since the belief in an old age for the Solar System is so entrenched, there has been a tremendous effort on the part of evolutionists to come up with explanations that would assure the Solar System of a continual (with time) supply of comets thus negating the argument for young age of the solar system. It would be profitable to examine these ideas and list the basic objections to them in order to see their worth.

1. Capture Hypothesis for Short-Period Comets
Many astronomers have believed that the short-period comets were captured gravitationally in a manner shown in (Fig. 4). As early as 1884 R. A. Proctor[14] pointed out that the theory of the capture of comets of short-period by the planets involved many serious difficulties. He argued that there would not be enough close approaches

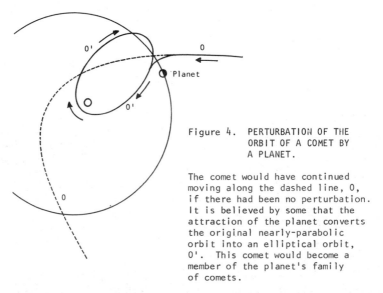

Figure 4. PERTURBATION OF THE
ORBIT OF A COMET BY
A PLANET.

The comet would have continued
moving along the dashed line, 0,
if there had been no perturbation.
It is believed by some that the
attraction of the planet converts
the original nearly-parabolic
orbit into an elliptical orbit,
0'. This comet would become a
member of the planet's family
of comets.

of long-period comets to Jupiter, for example, to provide it with such a large family of comets. Recently, comprehensive investigations by Edgar Everhart[15] confirm that the peculiarities in the observed distribution of short-period comet orbits cannot be explained on this basis at all. S. K. Vsekhsviatskii[16] has listed a number of basic arguments against the capture hypothesis of comets:

"(a) The number of known comets of the Jupiter family is 10^5 times as many as would be theoretically expected according to the capture theory. (b) The discrepancy between the actual and theoretical number of short-period comets compared to the number of comets with periods up to 9, 30, 100, and 1,000 years are 64, 74, 90 and 125, respectively, while according to the capture theory, these numbers should be 64, 310, 1,500, and 23,000. (c) Also, the absence of retrograde motions in the Jupiter family cannot be understood from the point of view of the capture theory. It was H. Newton who found that 30% of all short-period comets should, on the capture theory, have inclinations greater than 90°. According to Scigolev, half of all short-period comets should have retrograde motion. (d) The observed eccentricities of the short-period comets are often smaller than the minimum values predicted by the capture theory. The past few decades have seen the discovery of comets with almost circular orbits, which cannot be explained by capture."

Thus, it would seem that the presumed capture is unsupported by observation. The observed short-period comets could not have been formed from the field of long-period comets by means of single great perturbations or as successive small ones.

2. Comets from the Asteroids Hypothesis

Chamberlin suggested that comets might be produced by the disruption of small asteroids passing within the Roche limit of the great planets. The existence of over 1,500 asteroids already discovered suggests that very many more are present in the solar system either at greater distances than of those yet observed or having far smaller mass and size than them, but not yet discovered. Walter Baade has estimated that the number capable of discovery by use of the 100-inch telescope at Mt. Wilson exceeds 30,000, but as there may in fact be no definite lower limit of size, the actual number may be far larger than this.

Some of the objections to this hypothesis are: (1) A serious difficulty is met in explaining comets of high inclination moving in orbits that do not pass near that of any planet, and indeed, few comets at present have orbits that pass within anything approaching the Roche limit of Jupiter. Thus, the hypothesis, if it were admissible, might explain some comets, but not all. (2) Where the mechanism itself is concerned, however, break-up of an asteroid by gravitational action appears to be a dynamical possibility, but cohesive forces in the asteroids would place a serious lower limit to the general size of the resulting fragments. According to calculations by Jeffreys[17] an asteroid 200 km or less in radius could graze the surface of Jupiter without undergoing disruption owing to its internal cohesion. Even if this value were correct only in order of magnitude, the asteroid's disruption into small pieces is an utter impossibility. (3) Some have suggested that comets represent an earlier stage of development of asteroids. As mentioned, there are some 1,500 known asteroids, all moving in orbits of moderate inclination to the ecliptic and every one, without exception, pursuing direct motion. This alone makes the suggestion an improbable explanation, for this hypothesis would need many additional hypotheses to get from the randomly distributed cometary orbits to the regular arrangements of the asteroid orbits in order to give it even a vestige of tenability. (4) From the physical standpoint, there is no real resemblance at all between asteroids and comets, possibly apart from their small masses. (5) If any connection exists there should exist meteor streams associated with asteroids that come within Earth's orbit.

3. Volcanic Eruptions of Comets Hypothesis

Another hypothesis is that comets are expelled from the bodies of some planets. This idea, such as it is, seems to have been associated principally with the name of R. A. Proctor, though its originality and ingenuity are so evanescent that it can hardly have escaped anyone else until 1870, the time of its popularity. No doubt numerous astronomers before Proctor conceived the idea. S. K. Vsekhsviatskii[18] is the modern-day advocate of this position who insists that comets come from volcanoes operating on Jupiter. It would require an exceptionally uncritical attitude to overlook its apparent objections and weaknesses.

The objections are numerous: (1) We see that it is not a hypothesis in the scientific sense for it has no foundation either in observation or theory quite apart from the subsequent question whether or not it fits in with known properties of comets. If a single comet were ever to have been observed emerging from a planet, this would provide a valid basis for the idea; or if the theory of the structure of planets were to show that expulsion of part of their mass was possible in a form capable of being identified with a comet, this again would afford a reasonable basis for the idea and it would become a legitimate hypothesis. This is not the case, however. (2) If comets had emerged in some way from planets, the association of their orbits with the planetary orbits would be far stronger than it is, since apart from extraneous disturbances each cometary orbit would be extremely closely related to that of its parent planet. The few comets that are so associated represent a negligible proportion of the number that exist, but even for these the closest distances from Jupiter's orbit are very great compared with what they should be, at least for some, if the comets had actually been ejected from the planet. (3) Nearly parabolic orbits would be the exception and not the rule for such an origin. (4) If on the other hand we turn to the planets themselves, only Jupiter can by any extension of thought be reckoned as the source. But for small solid particles to be driven upwards through the vast atmosphere of this planet would require unreasonably high initial speeds, and we know that even in Earth's much shallower atmosphere meteors (believed by many to be parts of comets) are completely vaporized. This alone should make it clear that there is no possibility of anything remotely resembling a comet coming out of a planet. The requisite high speed would cause it to vaporize at once, and a jet of gas at high temperature fired out of Jupiter, even supposing it could penetrate the atmosphere, which is highly doubtful, would probably dissipate as soon as it got into the empty space beyond. Then again, many comets have far greater overall volume even than Jupiter, and a few greater than the sun itself, whereas one would expect that anything shot out from a planet would if anything be considerably smaller in size than the source. (5) What little is known of the internal structure of planets is against the idea of ejection of a swarm of dust particles. (6) The orbits of long-period comets show no special relation to that of Jupiter. (7) A comet to be born on Jupiter, must first get free of its surface. The speed of escape from Jupiter is 37 miles per second, about five times that for Earth. This is the escape speed as calculated from the observed equatorial diameter of the planet, which includes the atmosphere. The escape speed from the solid surface will be higher—how much higher depending upon the depth of the atmosphere. In the last thirty years various authorities have estimated the depth of the atmosphere as somewhere between 100 miles and 10,000 miles. If we assume the more likely value of 10,000 miles, the escape speed from the solid surface would be 42 miles per second. (8) Even if energy sources exist within Jupiter capable of imparting such a speed to a body, that alone is not necessarily going to make a comet out of the body. To be made into a comet, the body must first penetrate the atmosphere and then have enough speed left to achieve escape. We accomplish this feat on the earth by keeping a rocket in powered flight until it is through the bulk of the atmosphere at about 40 miles, after which it can safely be accelerated to escape velocity. But does a volcano have the technical know-how to launch a comet successfully? A volcano would give a body an initial powerful impulse, after which it is strictly on its own. Coslin has estimated that a particle at the surface of Jupiter would need an initial speed of 370 miles per second to get through the atmosphere and escape.

4. Oort's Cometary Cloud Hypothesis

In 1950 the Dutch astronomer Jan H. Oort[19] proposed that a vast swarm of comets (2×10^{11}), more numerous in number than the stars in the Milky Way, exists in a sperical shell extending from some 30,000 to 100,000 AU from the sun. The total mass of matter in this shell would be only 20 to 200 times the mass of the earth ($M_\oplus = 6 \times 10^{27}$ g). He supposed that occasionally one of the incipient comets leaves the shell, perhaps because gravitational perturbation of a nearby star has tugged it out of place, and the comet approaches the sun on a different orbit. Its orbit would now be a long ellipse (an ellipse of high eccentricity). The reasoning is that if the comet passes near the Jovian planets its orbit would be altered by the gravity of these massive objects and it eventually would become a short-period comet.

Although more than 500 long-period comets have been observed, only about 50 have fairly accurate orbits determined for them. It is claimed that the most recent aphelion-points for those long-period comets show a high spatial density within a certain radial range of large distance from the sun. The comets used for this study must be the ones for which planetary action has been taken into account since the osculating orbit based on the limited range of path near the sun over which the comet is observable may not yield even a fair approximation to the actual greatest distance from which the comet last came. Only about 10% of the known long-period comets fit this criterion to date.

Oort decided to plot in a histogram the number of comets against the reciprocal of the semi-major axis($1/a$) with the available comets grouped together in successive small ranges of $1/a$ of extent 50×10^{-6} AU^{-1}. The binding energy of the sun for a comet is proportional to $1/a$. In Oort's first paper, using only 20 comets, a strong peak in the distribution for the interval corresponding to the smallest values of $1/a$ ($0 \leq \frac{1}{a} \leq 50 \times 10^{-6}$ AU^{-1}) was claimed. This was thought to be a significant property of long-period comets as a whole. This peak led to the idea of a "shell of comets." It was inferred that the boundaries of the shell were from 50,000 to 150,000 AU from the sun.

In Oort's second paper,[20] using 40 comets, the distribution of comets in equal small successive ranges of 50×10^{-6} AU^{-1} in 1/a was again found to exhibit a strong peak for the smallest values of 1/a. In this paper it was inferred that the boundaries were from 30,000 to 100,000 AU from the sun. The maximum distribution was at 50,000 AU from the sun. Needless to say even if all the alleged 200 billion comets in the shell were assembled together at a distance of 30,000 to 100,000 AU they would be so dim as to be undetectable by existing optical equipment.

The histogram similar to the one Oort used is shown in (Fig. 5). The maximum height, the general width of this sharp maximum peak, and the sweep down precisely to the origin all lie completely to the lefthand side of the last available point whose abscissa is 0.2×10^{-4} AU^{-1}. It is not surprising that, since a long-period for a comet implies a large semi-major axis and small 1/a, a high proportion of the long-period comets would fall in this part of the diagram since for a range of 1/a from 0 to 50×10^{-6} AU^{-1} corresponds to r=2a from r=40,000 AU to r=∞ AU. The volume enclosed in this region would be a very large volume. As a matter of fact, for the range out to which the sun could hold these comets against the action of the rest of the Milky Way, the volume of this shell would be about 100 times larger than the volume between the inner boundary of the shell and the sun.

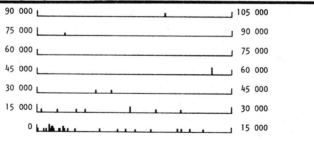

Figure 5.   ORIGINAL VALUES AT APHELION OF SEMI-MAJOR AXIS a (in AU) FOR 42 LONG-PERIOD COMETS.   (AFTER LYTTLETON)

There are numerous objections to this hypothesis. Some of these objections to Oort's shell model which have been advanced effectively by R. A. Lyttleton[21] are as follows:

a) Uncertainties of observation and calculation lead to random errors in 1/a of the order of 50×10^{-6} AU^{-1}. There is no way of knowing whether there may also be present a systematic error above or below this value. It would seem very questionable whether it is possible to assign a comet a position using equal intervals of 1/a with accuracy as fine as Oort used which was an interval of the same size as the order of error.

b) Large changes in 1/a occur as the comet moves in its orbit due to planetary action. This is believed to be of the order of 250×10^{-6} AU^{-1} or more. Thus it would seem impossible to know the original 1/a for the comet.

c) There seems to be a basic error in interpreting the meaning of the distribution plot using 1/a. If it is required to find the volume-density of some entity, then it is necessary to tabulate or graph numbers of that entity against some parameter directly represent-ative of that entity.

In this discussion I will follow Lyttleton. Suppose we have a popu-lation of objects (the long-period comets) such that in the range of y to y + dy of some associated numerically representable param-eter the number of members is N(y)dy.

If the particular parameter y is replaced by a relation $y = f(x)$, then the distribution of members in x to x+dx will be N[f(x)]f'(x)dx.

The number-density with x is N[f(x)]f'(x). There is no basis for asserting that a histogram peak of N(y) will necessarily imply one at the corresponding point in N[f(x)] f'(x). For the cometary case, $y = f(x) = 1/x = 1/a$, no distribution exists having this property at any finite value of a. For maximum of N(x) occurs at N(y) = 0, while a maximum of N[f(x)]f'(x) occurs where N[f(x)][f'(x)]^2+N[f(x)]f''(x)=0. This becomes $0 \cdot [f'(x)]^2 + N(y)2/x^3\}$ or $N(y) \cdot (2/x^3) = 0$. Since N(y)=0 at any finite point, more especially at a maximum, this cannot be satisfied for any finite x, that is, for any finite value a. No distribution whatever exists having this property at any finite value of a.

For example, take the first 500 integers, with respect to a real variable y. For this, the number-density N(y) would be constant; but if a plot against equal intervals of x=1/y were made, the numbers x to x +d x would be proportional to x^{-2}dx, and a histogram, using a very small interval $\triangle$ x, would show a strong peak for some small value of x. But obviously this cannot imply a congestion of integers surrounding the corresponding value of y.

To approach the matter in another way, let us suppose that there were actually a uniform volume-distribution of aphelion-points denoted by N per unit volume. Then the actual number to be found in the range a to a + da without regard to position on the celestial sphere would be proportional to a^2 da.

In order to arrive at the observed number it is necessary to notice that an adjustment must be made depending on the period P. A comet with a period of 100 years will appear in the inner reaches of the solar system 10 times for each apparition of one with a period of 1000 years. Thus a factor P^{-1}, or $a^{-3/2}$ must be multiplied into the actual number to arrive at the observed distribution in space, the observed numbers with regard to "a" alone should increase with "a" as fast as $a^{1/2}$.

If now we put b=1/a then plotted against b, the observed number would be proportional to $b^{-5/2}db$, and so even for uniform N a very strong peak, indicated by the factor $b^{-5/2}$ would occur near b=0. Any plot of numbers of long-period comets against 1/a will automatically exhibit a peak at small values of this parameter and cannot be inverted to demonstrate a high-volume density of aphelia in space.

d) When a plot is made of numbers of comets versus "a" itself, there is no sign of congestion at large values. Of the 54 comets used, 28 have aphelion distance less than the inner radius of the shell, only 10 within the shell while 16 are far beyond the outer radius of the alleged high-density shell.

Thus, we may conclude that the observational material used to provide a basis for a shell of comets when properly used establishes the nonexistence of the shell. The positions of actual aphelion points show no sign of any concentration at any range.

e) If comets were perturbed from almost circular motion in the supposed shell to the observed almost parabolic motions, the requisite changes in energy would in general displace the new aphelion points far outside the shell.

f) To maintain the supply of observable long-period comets by means of stellar perturbations from the shell would require hundreds of stars currently within a distance of the order of 1 parsec from the sun. (1 parsec = 3.3 light years).

5. Accretion Hypothesis for Comets

It has been proposed that comets are forming by gravitational accretion in meteor streams which allegedly come from the break-up and disruption of comets. This is very difficult to understand in the light of the perturbing influence of the planets and the Poynting-Robertson effect on particles. Further, this would go opposite to the direction indicated by the entropy law representing the direction of processes in this universe.

E. Conclusions Regarding Short-Period Comets

The failure to find a mechanism to resupply comets or to form new comets would seem to lead to the conclusion that the age of the comets and hence the solar system is quite young, on the order of just several thousand years at most. Occam's razor should be followed in this matter. The obvious is that the solar system has been operating on a short time scale since its creation.

Chapter IX

THE POYNTING-ROBERTSON EFFECT

A. Physics of the Poynting-Robertson Effect

The Poynting-Robertson effect[22] strongly perturbs the orbits of small orbiting particles (of centimeter dimensions and smaller) producing a tendency for the particles to spiral inward toward the sun. The angular momentum of revolution of an orbiting particle is progressively destroyed since it receives solar radiation which has only radial momentum from the sun (neglecting solar rotation) and re-radiates this energy with a forward momentum corresponding to its own motion about the sun. This is the essential feature of the Poynting-Robertson effect.

To look at the effect simply, consider a small particle moving in a circular orbit around the sun. Sunlight, which can be thought of as a steady stream of photons, flows outward from the sun while the particle moves at right angles to the photon stream. Thus, the photons strike the particles preferentially on its leading side, just as a car driven through a rainstorm is struck on its front side even though the rain may be falling vertically. The effect is shown in Figure 6. The apparent displacement of the source of the light is called the aberration of light. The Poynting-Robertson effect is caused by "drag" due to "apparent" displacement of direction of flow of photons from the solar direction since there would be two components of this apparent motion: one, parallel but opposite to the motion of the particle, and the other, perpendicular to the particle's motion and outward.

The preferential absorption of photon momentum on the leading side is only one part of a complex effect whereby re-radiation of absorbed photons causes a net loss of energy and the particle

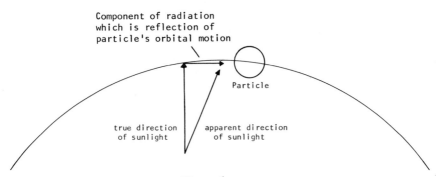

Figure 6.

slowly settles into a smaller and smaller orbit, spiraling in toward the sun. This effect was first predicted by the British physicist Poynting (1903) and was later amended and analytically verified by the American physicist Robertson (1937).

Consider a grain of dust orbiting about the sun in interplanetary space. It absorbs sunlight, and re-emits this energy isotropically. We can view this two-step process from two different standpoints.

1. As seen from the sun, the particle absorbs light coming radially from the sun and re-emits it isotropically in its own rest frame. A re-emitted photon carries off angular momentum proportional (i) to its equivalent mass ($h \nu/c^2$), (ii) to the velocity of the grain ($R \dot{\theta}$), and (iii) to the grain's distance from the sun (R). In these equations ν is the frequency of the radiation, h is Planck's constant, $\dot{\theta}$ is the angular velocity of the grain about the sun, and c is the speed of light in a vacuum. The grain loses orbital angular momentum L about the sun at a rate of approximately

$$dL = \frac{h}{c^2} \nu \dot{\theta} R^2 \ \text{OR} \ (1/L)dL = \frac{h\nu}{mc^2}$$

for each photon whose energy is absorbed and re-emitted, or isotropically scattered in the grain's rest-frame. The grain's mass is m.

2. Seen from the grain, radiation from the sun comes in at an aberration angle, $90° - \theta$ from the direction of motion of the grain. If V is the speed of the particle in a circular solar orbit, the angle (θ) between the incoming radiation and the radius vector to the sun is $\theta \approx V/c$. Here V is $\dot{\theta} R$, the grain's orbital velocity. Hence the photon

imparts an angular momentum $pR\sin\theta = \dfrac{h\nu}{c^2} R^2 \dot\theta$ to the grain where p is the momentum of the photon.

For a grain with cross section σ,

$$dL/dt = \frac{-L_\odot}{4\pi R^2} \frac{\sigma L}{mc^2}$$

where $L_\odot$ is the solar luminosity.

In either approach the grain velocity decreases on just absorbing the light. From the first viewpoint because the grain has gained mass which it then loses on re-emission; from the second, because of the momentum transfer.

3. A third way of arriving at the Poynting-Robertson effect on a particle and the time of its spiralling inward to the sun may be obtained rather simply in the case of circular orbits in the following manner.

Consider the special case of a spherical particle of mass m and diameter d in a circular orbit of radius r. Its orbital speed may be found from the relation:

$$(\text{Centripetal Force}) \quad = \quad (\text{Gravitational Force})$$

$$\frac{mv^2}{r} = m\frac{GM}{r^2} \,;$$

$$v = \left(\frac{GM}{r}\right)^{1/2}$$

M is the mass of the Sun and G is the gravitational constant. The total orbital energy is E = (Potential Energy) + (Kinetic Energy),

$$E = \frac{-GMm}{r} + \frac{1}{2} mv^2 = \frac{-GMm}{2r}$$

Let us consider separately the processes of absorption and re-radiation of energy.

In time dt the particle receives energy $d\varepsilon$ as solar radiation. This causes an increase in mass $dm = \dfrac{d\varepsilon}{c^2}$. But since this radiation traveled radially from the sun, it carried no orbital angular momentum and the total orbital angular momentum of the particle is conserved so that H = mvr = constant;

$$dH = md(vr) + vr\,dm = 0\,; \quad md(vr) = -vr\,dm = \frac{-v}{c^2} r\,d\varepsilon.$$

The particle then reradiates the energy $d\varepsilon$, but it does so isotropically in its own frame of reference and this process involves no reaction on the particle. The orbital velocity is therefore conserved in the radiation process and since the mass dm is lost, a net loss of angular momentum by v r dm occurs. This angular momentum is carried away by the radiation which, when viewed in the stationary reference of the sun, is seen to be Doppler-shifted; the energy and momentum projected forward from the particle exceed the energy and momentum radiated backward.

The rate of loss of orbital radial angular momentum may be equated to a retarding torque

$$\tau : \quad \tau = \frac{md(vr)}{dt} = \frac{-v}{c^2} r \frac{d\varepsilon}{dt} \quad \text{so that } \frac{dE}{dt} = \tau \frac{v}{r} = \frac{-v^2}{c^2} \frac{d\varepsilon}{dt}$$

Now $d\varepsilon/dt$ is the rate at which the particle receives solar radiation and is given by

$$d\varepsilon/dt = S\left(\frac{r_E}{r}\right)^2 A$$

where S is the solar constant or the energy flux normally through unit area at the distance r_E of the earth's orbit (S = 1.39×10^6 ergs/cm^2-s), and A = $(\pi/4)d^2$ is the cross-sectional area of the particle. By differentiating E with respect to t and equating this to

$$dE/dt = \frac{-v}{2} \frac{d\varepsilon}{dt}$$

we obtain the relation $\frac{GM}{2r^2} \frac{dr}{dt} = \frac{-v^2}{c^2} S\left(\frac{r_E}{r}\right)^2 A.$

But $v = \left(\frac{Gm}{r}\right)^{1/2}$ so that we get $r\frac{dr}{dt} = -\frac{2S r_E^2}{Mc^2} A.$

Integrating from the initial condition, $r = r_0$ at $t = 0$;

$$\frac{r_0^2 - r^2}{r_E^2} = \frac{4SA}{Mc^2} t$$ which, for a spherical particle of

density ρ, becomes $\left(\dfrac{r_o}{r_E}\right)^2 - \left(\dfrac{r}{r_E}\right)^2 = \dfrac{6s}{d\rho c^2}\, t$

where r/r_E is the radius of a particle orbit expressed in astronomical units.

Table II lists the names of falls for particles of various sizes and orbits calculated with more complete formulas which include the eccentricity.[23] The Poynting-Robertson effect would seem to be quite efficient in rapidly clearing the solar system of dust material. But the solar system is quite "cluttered" with dust and gas material.

TABLE II
(after Wyatt and Whipple)
Times of fall for particles with various orbits

a, AU	e (eccentricity)	$t/(10^7 \text{ yr} \times d/2\rho)$	Type of Orbit
3	0.0	6	asteroidal
3	0.7	0.7	asteroidal
1.4 (Geminids)	0.9	0.14	meteor showers
3.5 (Bielids)	0.7	3	meteor showers
10 (Leonids)	0.9	7	meteor showers
55 (Lyrids)	0.98	20	meteor showers

The Poynting-Robertson effect causes a general reduction in the eccentricity and major axes of small bodies in orbit about the sun until, other circumstances permitting, they spiral near the sun to sublimate. The time of spiraling into the sun for a black, or perfectly reflecting, spherical particle of radius d/2 and density ρ in an orbit of perihelion distance q(in AU) is given by the following equation:

$$\text{spiral time} = C(e)\,\frac{\rho d q}{2} \times 10^7 \text{ yr}$$

where the factor C(e) depends solely upon the orbital eccentricity e.

B. Some Calculations of Spiral Times of Particles in Circular Paths

To show some of the times for particles of different sizes to spiral into the sun from initial circular orbits about the sun we can use the equation

$$t = 7 \times 10^6 \, \frac{\rho d r_o^2}{2}$$

(years) where the particle radius d/2 is in centimeters, the particle density ρ is in gm/cm³, and the particle distance r_o is in astronomical units. Assume a particle density of 2.7 gm/cm³ for these calculations. Table III lists the results of these computations.

TABLE III

Times (years) for spiralling into the sun for particles of different radii and at different distances from the sun

radii (cm)	distance (AU) 0.5	1	3
10^{-5}	23.6	189	1,701
10^{-4}	236	1,890	17,010
10^{-3}	2,363	18,900	170,100
10^{-2}	23,630	189,000	1,701,000
10^{-1}	236,300	1,890,000	17,010,000
1	2,363,000	18,900,000	170,100,000
10	23,630,000	189,000,000	1,701,000,000

It would appear that the solar system should be swept clean of particulate matter if it is very old. The typical lifetime for interplanetary dust particles is of the order of 10,000 years. Since the solar system seems to have an interplanetary matrix of small particulate matter, the age of the solar system cannot exceed 10,000 years.

C. Dispersion of Meteor Showers

There is a basic difficulty in trying to arrive at an actual age for the solar system using the results to be expected from the Poynting-Robertson effect, but it is certainly possible to put an upper limit on this age. This is evident regarding the dispersion of meteor streams that should take place by the Poynting-Robertson effect.

Meteor showers are believed to be debris moving in definite orbits. In many cases, the debris is distributed uniformly around the orbit. As regards the shower meteors, it seems natural to assume that the debris originally existed in one conglomerate—such as a comet—and that the dispersive forces have scattered the debris in the course of time around and about the orbit, though the proof certainly is circumstantial. The effects bringing about this dispersion are mainly planetary perturbations, ejection of meteoric material from comets, and the Poynting-Robertson effect. The extent of the dispersion gives possibilities of estimating the maximum age of the showers and, consequently, of the solar system, itself.

The lifetime of small meteoric bodies in the solar system is not at all in the range demanded by the evolutionist for the age of the solar system. For example, a particle of radius 0.05 cm and density

$4gm/cm^3$, corresponding to about a 5th magnitude meteor moving in the orbit of Halley's Comet, would be drawn into the sun in about 10 million years. Taking similar particle in the Lyrids associated with Comet 1861 I, it has been calculated that the time to sweep this particle into the sun would be about 37 million years. Now this is for a particle whose orbit extends beyond Neptune. It is difficult to see how there could be any meteoric matter at all in the solar system because of the very rapid break-up of comets and the Poynting-Robertson effect which acts continually upon both the particles making up the comets and the debris left from the comets themselves if the solar system is very old.

A further factor of great interest in this aspect of the age of the solar system is the possibility of observing dispersion in the meteor stream because of the Poynting-Robertson effect in the major showers by virtue of the selective effect arising from differing particle size and density—the smaller ones being drawn towards the sun faster than the larger ones. Wyatt and Whipple[24] estimated the order of times required to separate meteors of magnitude +5 and –2 by the Poynting-Robertson effect so that the earth would pass from one limit to the other in five days as it crossed the meteor stream. In order to do this, the showers in Table II with inclinations less than 40° were considered, and their inclinations assumed to be zero. The perihelion advance of the orbit was neglected and the earth's orbit taken as circular. The density of the individual meteors in each shower was taken as $4gm\ cm^{-3}$ and the radii of the meteors were calculated from the luminosity-mass relationship. The times for this sort of separation are vastly smaller than evolutionist estimates of the age of the solar system.

Wyatt and Whipple[25] state that such a separation has not been observed photographically. However, Lovell[26] says that in the case of the Geminid shower the radio-echo data show that there is a very marked separation of particle sizes. The most active region of the shower is near the end and consists of a marked concentration of heavier particles compared with the earlier activity of the shower. In other words, the dispersion is of the type to be expected from the operation of the Poynting-Robertson effect over a period of around 10,000 years. As regards the other showers, no separation is found, a result to be expected in view of the showers' recent origin. Perturbative effects other than the Poynting-Robertson effect are non-selective as regards particle sizes. What little dispersion that may have occurred may have been due mainly to planetary perturbations.

It seems that planetary perturbations, coupled with the possibility or differential ejection velocities from the original conglomerate, are the main influences dispersing meteoric debris around the orbit. The Poynting-Robertson effect introduces a selective drag which in the course of time will separate out the particles in a shower according to their size, eventually causing all to fall into the sun— the small ones faster than the large ones. This separation does not seem to have occurred except for the Geminids which may be explained mainly by planetary perturbations. This would indicate only several thousand years have elapsed since the origin of the streams at most and, thus, indicating a short age for the solar system.

Other effects operate to bring about a rapid destruction of meteor streams. In the eighteenth and nineteenth centuries great meteor streams were found associated with Biela's Comet. After 1899, however, the shower almost completely disappeared. This disappearance is coupled with the strange disruption of Biela's Comet. In more recent times a strong shower associated with the Pons-Winnecke Comet has also disappeared.

It is not at all established that all meteor streams are associated with comets. There are only six meteor streams that have received universal recognition as being associated with comets. There are a number of major showers for which there seems to be no possibility of a cometary relationship. There are some sixty-eight cases of of long period or parabolic comets which should approach the earth sufficiently closely to produce a meteor stream with any of these. The failure to find meteor streams associated with the majority of the elliptical comets and for sixty-eight parabolic comets with equally close approaches, and the existence of several major showers without parent comets certainly would say that most meteors may not be debris of comets. Porter prefers a common origin of a comet and the associated stream to the shower's formation from the comet. He insists that the May Aquarids and Orionids could not have originated from Halley's Comet in its present orbit for they are observed quite far from the orbit. The main features of an old meteor stream appear to be a duration of many days or weeks, a large dispersion of the radiant, and small changes in the hourly rates from year to year. It is believed that the showers can attain this stage after a time which is astronomically short. It has been estimated by Ahnert-Rohlfs and Hamid that the age of the Perseids is on the order of 10^4 to 10^5 years at the maximum. In the light of practically no dispersion of the streams by the Poynting-Robertson effect, the streams must be young.

A good example of showers definitely associated with a comet are Taurid-Arietid streams from Encke's Comet. Whipple and Hamid conclude that the streams were formed by the violent ejection of material from the comet some 5,000 to 14,000 years ago.

D. Sporadic Meteors

Another important aspect of this problem involves the source of sporadic meteors (those not associated with known showers). Some debate has occurred whether these meteors are the debris of the primeval matter from which the solar system was formed or whether they are the result of some subsequent cometary break-up. In this connection it is interesting to notice that the Poynting-Robertson effect sets a short-time scale at which any such break-up could have occurred. The mean semi-major axis for the asteroids is about 3AU and if the sporadic meteors were formed by a planetary disintegration at about this distance from the sun, their orbits would mostly be within that of Jupiter with an apse near that planet's orbit. Wyatt and Whipple have calculated the times of fall into the sun of particles with radius r moving in orbits of this type assuming their density to be $4gm/cm^3$. Their results show that for a time-scale of 3 billion years all bodies with a radius of less than 4 cm must have been swept into the sun. Again we have an upper limit on the age of interplanetary matter that is below that of the alleged age of the solar system advanced by the evolutionist. If a planetary break-up occurred sixty million years ago as some believe, then all particles originating with the asteroids and of radius less than 0.08 cm must have been lost into the sun. Thus, all meteoric material would have to be bright regardless of the two models adopted. There should be nothing fainter than about a fifth magnitude meteor. But there are numerous meteors continually seen fainter than fifth magnitude. They are not from interstellar space. The true ages of these meteors would seem to be much smaller than their alleged age from the evolutionist viewpoint.

Further, with the production of meteorites from an asteroidal origin directly as fragments of colliding asteroids in the asteroid belt or from those few asteroids which happen to cross the orbit of Mars, several problems rear their heads. The major and seemingly fatal objection to this idea is that the asteroids are too few to yield sufficient debris through collisions. Also, it seems that neither asteroids nor comets are sufficiently massive to lead to compositional stratification, so the irons could not have come from asteroids.[27] Whether sporadic meteors are the debris of the primeval matter from which the solar system was made, or whether they are the result of some subsequent planetary break-up, is a major problem yet to be solved. The evidence seems to lean to the position that the sporadic meteors are remains of primeval matter.

Öpik[28] has studied the effect of filtration of the particles through collisions with the planets. According to Öpik's calculations, Jupiter presents a major obstacle, and within the alleged age of the solar system, will have cut off practically all particles over 2mm in diameter, although letting through fairly well those of 0.2mm diameter. If this view is correct, then no meteor within the visual and normal telescopic range of observation can exist if the solar system is old and the meteors are remnants of primordial material. There is an infinite number of meteors in existence apparently. The solar system must be quite young.

E. Gases Surrounding Hot Stars

In considering the Poynting-Robertson effect in regions external to the solar system, the time for a body in a given orbit to spiral into a star will be equal to that for the same body in the same orbit around the sun multiplied by the factor E_o/E, the ratio of the total energy per second emitted by the sun to that emitted by the parent-star of the particle. For particles near the large, hot B stars the times will be shorter than those for particles near the sun by a factor of about 100,000.

There are many O and B stars surrounded by dust and gas clouds today. How could they be anything but very young stars? If the stars are young, it would seem that the universe of galaxies must be young also.

Chapter X

CHANGES IN SATURN'S RINGS

A. Nature of the Rings of Saturn

Saturn is a breathtaking sight when seen with a telescope, for it is unique in being one of the few planets with rings. Surrounding the planet and concentric with it is an extensive but very thin disk-shaped structure, lying precisely in the plane of Saturn's equator. Its shape is similar to that of a deflated inner tube. The rings exhibit varying degrees of brightness, with dark gaps.

From the moderately bright outer ring (ring A) we cross the dark Cassini division inward to the very bright middle ring (B), then across a smaller gap to the dim crepe ring (C), and finally to the newly-discovered, nearly-invisible subcrepe rings (Figure 7). The total radius of the ring system is about 2.3 times the equatorial radius of Saturn. The rings are separated from each other by gaps or divisions. From the earth, when the rings are viewed edge-on they

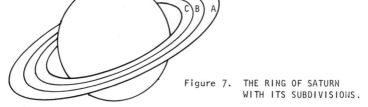

Figure 7.   THE RING OF SATURN
            WITH ITS SUBDIVISIONS.

B is the brightest part of the ring.
It is separated from the slightly
darker ring A by Cassini's division.
C is the crepe ring.

are practically invisible, while about 7½ years later they are inclined about 27° to the line of sight and are spectacular.

Galileo first observed Saturn's rings with the telescope in 1610, but Christian Huygens was the first to explain their true form in 1655. The physical nature of the ring system follows from these facts: (a) stars may be seen through the rings, so they are not a solid body; (b) the rings shine simply by reflected sunlight with no absorption features characteristic of gases, so they are composed of solid particles akin to the meteors or meteorites (some believe they are a mixture of "ice" and rock); (c) spectra of the rings show motional Doppler-shifts consistent with these particles being in Keplerian (the inner parts move with higher speeds than the outer parts) orbits about Saturn, though with speeds nearly that of particles in circular orbits at the same distance; and (d) the thickness of the rings is less than 10 km. The sizes of the particles are probably between 10^{-4} cm and several meters, and they seem to be composed of ammonia and/or water ice and rock.

B. Origin of the Rings

One view of Saturn's rings is that they are material brought into existence at the creation of Saturn but not used to make an ordinary satellite or form part of the planet. Certainly it is very doubtful that by a process of agglomeration the matter could have formed a satellite.

The tidal force exerted by a parent planet decreases as the inverse cube of the distance from the planet. Thus Saturn's tidal force is weak at a large distance and is unable to prevent matter from drifting together. But inside a spherical zone with radius 2.5 times that of the parent planet, the tidal force exceeds the mutual gravitational force of two moonlets in contact preventing them from staying together. So another commonly held view is that the rings were formed when a former satellite of Saturn ventured too close to its central planet and was torn to pieces by its attraction. Whichever of these two views is correct has a strong bearing on the age of the rings and the age of Saturn.

Work by the British geophysicist/astronomer Harold Jeffreys seems to indicate that the origin of the rings by disruption of an asteroidal-like object or a satellite in a near approach to Saturn is an utter impossibility. Roche's argument of the existence of a critical distance for a satellite within which it would be broken up by tidal action assumed that the satellite was in a fluid state. It was argued by Roche that a small liquid satellite moving in a circular orbit about a primary, the periods of rotation and revolution being equal, would take an approximately ellipsoidal form provided that it was not too close to the primary. If, however, the mean distance was less than about twice the radius of the primary (depending somewhat on the ratio of the densities) there would be no permanent form and the satellite would be broken up. Extensions of Roche's argument have been given by George Darwin and Sir James Jeans. The theory has been given for gaseous satellites and the restrictions that the satellite is to be small has been removed. On the whole, the modifications give surprisingly little change in Roche's main result. But the hypotheses all suppose the satellite to be fluid, and the corresponding problem for a solid satellite seems to have escaped an explicit solution. Jeffreys has derived the solution for a solid satellite from a study of the straining of an elastic sphere. I shall follow Jeffrey's approach[29] to this problem.

If the body is at a distance c from the center of mass of the planet of mass M and rotates with an angular velocity ω in a plane through the radius vector the deformation can be derived from a gravitational potential

$$V = \frac{GM}{2c^3}(2x^2 - y^2 - z^2) + \tfrac{1}{6}\omega^2(x^2 + y^2 - 2z^2).$$

Jeffreys considered the case of a satellite that is being made to approach its primary by tidal friction or a resisting medium to test the hypothesis that the rings came from rupture of a moon of Saturn. So for a satellite gradually approaching its primary and, finally near the primary's surface, it is found that the greatest stress difference at the center of the small body is

$$(32/19)\ \frac{GM\rho}{c^3}a^2 \qquad \text{where}$$

G is the gravitational constant, M is the mass of the primary, c is the distance from the center-of mass of the primary to the moon, a is the radius of the satellite, and ρ is the density of the satellite. If ρ_0 is the density of the primary, we obtain $M/c^3 = (4/3)\pi\rho_0$. Since c is approximately the radius of the primary the previous relation says that for rupture to occur

$$(4/3)\pi\frac{(32)}{19}\ G\rho_0\rho a^2 > 10^9.$$ The 10^9 is the

magnitude of the critical stress difference for the material composing the satellite.

Suppose a sphere of ice approached Saturn and comes to the mean distance of the rings of Saturn (c $\approx 10^{10}$cm). Using the equation

$$\frac{32}{19}\left(M/c^3\right)G\rho a^2$$

and taking $M = 6 \times 10^{29}$ gm, $c = 10^{10}$ cm, $G = 6.67 \times 10^{-8}$ dyne – cm^2/gm^2, $\rho = 1gm/cm^3$, $\rho_o = 0.7$ gm/cm^3, and the crushing strength of ice as 3×10^6 dyne/cm^2, we get a > 133 km or the diameter of the moon must have been greater than 266 km. Hence, if an ice satellite ever revolved about Saturn at the mean distance of the rings, and was broken up by tidal action, its diameter was over 266 km; a body smaller than this would not be disrupted. Since the rings appear equally bright all around and its maximum possible thickness is given as 10 km, it would seem that we must conclude that the rings were not formed by disruption of a solid satellite. Thus, the first mentioned hypothesis regarding the origin of the rings must be correct and the age of the rings is the same as the age of the planet.

C. Stability of Saturn's Rings

The precise circularity of the orbits and their extreme confinement to the equatorial plane are probably explained by mutual collisions of the particles. If, in the past, individual orbits were inclined to one another, then ultimately each up-bound object would run into a down-bound object; on the average reducing the orbital inclinations. If the orbits were at one time eccentric, each out-bound object would finally hit an in-bound one; on the average reducing the eccentricities. When the process had continued for sufficient time, the state of affairs would become much as we see today. The varying brightness of the rings with distance from the planet indicates differences in the number density of moonlets, the variations result principally from the original distribution of the material and the perturbing effects of Saturn's satellites. The period of an object in Cassini's division is approximately two-thirds that of Janus, one-half that of the next satellite, Mimas, one-third that of Enceladus and one-fourth that of Tethys. The perturbing gravitational pull of Mimas will be in the same direction in space every 0.94 day when a moonlet in Cassini's gap or division has Mimas at opposition and at its minimum distance. The cumulative effect of the satellites is to remove objects moving in this zone and to place them into orbits of a different size. Other less pronounced gaps in the ring system correspond to other simple fractions of the orbital periods of the inner satellites.

Harold Jeffreys[30] has shown in another classic paper that if Saturn's rings were several particles thick to start with,the damping effect of collisions on the inclinations and eccentricities would reduce the rings in less than a year to a state where the particles were piled one on another. Dissipation by friction and impact would not cease at this stage. Its later effects would extend the ring inwards and outwards in its own plane until it was nowhere more than one particle thick and the spacing was just enough for collisions to be avoided. The upper limit of time needed to attain this state is estimated to be of the order of 10^6 years for particles of diameter 1 cm and less for larger ones. Stability of the rings then is achieved quite rapidly, relatively speaking. Obviously, if the rings are presently changing toward this stable state discussed by Jeffreys then the age of the rings and Saturn would have to be less than a million years.

In this discussion I will follow closely the approach of Jeffreys. The high opacity of the rings shows that on an average most rays of light striking the ring at angles up to 27° meet a particle on their way. Jeffreys argues that every particle of the ring must cross the mean plane of the ring twice in each revolution. There should be two collisions of the particle on the way. The point Jeffreys makes is that the opacity and frequency of collisions depend on the same function of the number and size of the particles. The total surface of the particles from steady motion in circles can be treated as random and the opacity shows that collisions will have a dominating effect. Collisions between solids are ordinarily non-conservative. For each collision the relative velocity between particles is reduced because of friction and imperfect restitution by about 1/2. The relative velocities between particles will be rapidly cancelled, with this time being not more than an orbital period, approximately a day. Jeffreys believes that in the case of particles originally in the anchor ring, for an average normal to the plane of revolution intersecting several particles, the velocities normal to this plane would be practically destroyed in a year. It would seem that the radial velocities would be removed at the same rate. The ring would become thin, but at any distance several particles would be piled one on another to stay. Jeffreys argues convincingly that this state could last a long time with little change because the particles would acquire such rotations that there would be little difference of velocity at the points of contact. Dissipation would not be completely done away with.

The major cause of further dissipation would be the differences in orbital period. Let us consider the physics of this problem as sketched by Jeffreys. Take the particles as spheres of radius a, and consider two sets moving in circles of radii r, r+b, where b < 2a. By Kepler's third law we could write the ratio of the periods of the particles as $(r/r+b)^{3/2} \approx 1-3/2\ b/r$. Thus if the spacing in longitude of the particles is also b, a particle on the inner circle will encounter on an average 3/2 particles in a revolution and share momentum with them. If v is the orbital velocity at a distance r; m, the mass of a particle; the outward transfer of angular momentum at a collision is of the order

$$-mbr\left(\frac{dv}{dr}\right),$$ the numerical factor being less

than 1 but probably more than 0.1. A particle transfers outwards, per unit time, an angular momentum of order

$$\frac{-3mbv}{4\pi}\frac{dr}{dr}.$$ The

particle occupies an area 2ab. This transfer of angular momentum per unit time is equivalent to a tangential stress

$$-\lambda\frac{mv}{ar}\frac{dv}{dr}$$ where λ

will probably be between 0.01 and 0.1. Since $v^2 = \mathcal{U}/r$, the tangential stress is

$$-\lambda\ m\mu/2ar^3,$$ and for the total rate of outward transfer of

angular momentum in a belt where the normal section is

$$\pi\lambda\mu cm/ar^2.$$

The total angular momentum in a ring, if a, b, c are constant, is

$$\tfrac{1}{3}\mu^{1/2}\frac{cm}{ab}[r^2].$$

Comparing these results, the time that would be needed to transfer the whole of the angular momentum from the inner to the outer half of the mass would be of the order

$$(2/3\pi\lambda)\frac{r^{1/2}[r^{1/2}]}{nb}$$ where n

is the mean motion of a particle at the mean distance of the rings.